DENIZENS OF THE MOUNTAINS

SIERRA NEVADA. TYPICAL NUTCRACKER COUNTRY

DENIZENS OF THE MOUNTAINS

BY
EDMUND C. JAEGER
AUTHOR OF
THE MOUNTAIN TREES OF SOUTHERN CALIFORNIA
DENIZENS OF THE DESERT

ILLUSTRATED BY PHOTOGRAPHS

1929
CHARLES C THOMAS · PUBLISHER
SPRINGFIELD, ILLINOIS   BALTIMORE, MARYLAND

Copyright 1929 By
Charles C Thomas
Manufactured In
The United States

Nature Books

We offer in *Denizens of the Mountains* the introductory volume of *Nature Books*, to be followed soon by others of interest and importance. *Nature Books* have been established to disperse authentic natural history in many aspects in a dignified, readable, and appealing form. The field to be covered is broad. Not only will there be books on birds and mammals, but such varied natural objects as stars, clouds, rocks, trees, flowers, insects, fish, in fact all nature, fall within the intent of these volumes. Geographic boundaries offer no limiting factor, and the subject matter may vary from purely descriptive to methods, biography, exploration, philosophy. Insistence will be stressed that the writer of each volume be an authority on the subject covered, that the matter be presented in an entertaining literary style, and that the text be amply illuminated with illustrations of high grade. We plan to be productive in a neglected field of natural history and American literature, and we have confidence that our readers will profit by the results of our efforts.

Hartley H. T. Jackson,
Editor

AND WHILE THE AROMA OF THE GREASE-WOOD WAS STILL ON THE AIR, I WAS ON MY WAY BOUND FOR THE MOUNTAINS

To
S. STILLMAN BERRY
inspirer of youth
and
five faithful friends
of the trail
GREGORY D. HITCHCOCK
RUSSEL BAILEY
KENNETH MAXWELL
PHILLIP SAVAGE
RICHARD JONES

PREFACE

This book constitutes a series of life history sketches of some of the more familiar and interesting mammals and birds of the mountains of the western United States. The material on which these studies is based has been gleaned from experiences with the wild creatures in their native homes in the Rockies, Sierra Nevadas, and intervening ranges of Utah and Arizona. The author hopes that his narrations show because of this an element of freshness that will make them attractive not only to those who travel often the trails of the highlands, and who have therefore had similar experiences in the field, but also to those who seek to learn for the first time the habits and mannerisms of their native fauna. It is the author's further desire that these sketches may be helpful to the teachers in our schools who may wish to place in the hands of their pupils familiar studies, which, while deepening their friendship with the wild life of their environment, will contribute directly to their knowledge of natural history. Every effort has been made to give only dependable information and to avoid ascribing to the lower creatures mental states that they do not possess. By keeping somewhat the specialist's point of view before the reader it is hoped that he will learn to interpret the facts of animal behavior in a proper way and be prepared to take up later the more serious study of technical papers.

During the preparation of this manuscript the author has sought the guidance and help of specialists in his field and he wishes to acknowledge with gratitude the many courtesies they have extended to him. Special thanks are due to A. Brazier Howell of Johns Hopkins Medical School, who has had the kindness to read many of the chapters and make needed suggestions, and to Wright Pierce of Pomona College, and Donald R. Dickey of the California Institute of Technology, who have furnished many of the photographs. Richard Allman of the Riverside Junior College contributed the line drawings. Public acknowledgement is made of the encouragement tendered the author by Dr. Loye Holmes Miller of the University of California and Dr. Harold C. Bryant of the California Fish and Game Commission.

Edmund C. Jaeger

Riverside Junior College,
Riverside, California,
July 2, 1929.

CONTENTS

	Foreword	v
	Preface	ix
	List of Illustrations	xiii

CHAPTER		PAGE
I	The Mountain Wood Rat or Bushy-Tailed Neotoma	1
II	The Cony	9
III	The San Bernardino Chipmunk	17
IV	The Merriam Chipmunk	31
V	The Mountain Chickadees	39
VI	The Anthony Gray Squirrel	44
VII	The Mountain Weasel	56
VIII	The Nuthatches	62
IX	The White-Footed Mouse	67
X	The Clarke Nutcracker	77
XI	The Belding Ground Squirrel	85
XII	The Blue-Fronted Jay	94
XIII	The Gray Fox	104
XIV	The Water Ouzel or Dipper	112
XV	The Golden-Mantled Ground Squirrel	122
XVI	The Wren-Tit	131
XVII	The Western Striped Skunk	137
XVIII	The Junco or Snow Bird	145
XIX	The Mutillids or Cow-Killers	152
	Index	163

ILLUSTRATIONS

Bound for the Mountains . . . Facing Dedication Page	
Sierra Nevada . . . . . Facing Title Page	
Female Mule Deer	2
Map Showing Distribution of Subspecies of Bushy-Tailed Wood Rats	6
In the Debris at the Bases of the Cliffs the Conies Live	10
Sierra Cony	11
San Bernardino Chipmunk	19
Chinquapin Thicket: Home of the Chipmunk . .	20
Pancakes for Breakfast	25
Anthony Gray Squirrel	45
When Snows are Deep Even Gray Squirrels Sometimes Attack the Bark of the Pines	54
Wild Mice Are Common Forest Neighbors . . .	72
Clarke Nutcracker	78
Grouse Meadows. High Sierra Home of the Belding Ground Squirrel	86
The Belding Ground Squirrel	87
A Ground Squirrel Meadow	93
The Inquisitive Blue-Fronted or Stellar Jay . . .	95
Young Stellar Jays	102
Arizona Gray Fox	107
The Home of the Ouzel	113
Nest of the Ouzel	119
The Golden Mantle or Calico Squirrel . . .	123
The Western Porcupine	126
Pallid Wren-Tit	132
Thurber Junco—near nest	146
Nest of Thurber Junco	148
The Mutillid	156

CHAPTER I

THE MOUNTAIN WOOD RAT OR BUSHY-TAILED NEOTOMA

4:00 A.M.

THE yellow gibbous moon had set behind the canyon wall. Peering through the trees on the skyline it went to rest, flooding the great forest of sombre pines with its mellow light. Soon a dawn breeze began to blow lightly over the face of the meadow and with it came the fresh aroma of dew laden leaves. The blue-fronted jays, aware of the coming of day, were shaking their crests and coming down from their perches up among the oaks. Small notes from the sparrows announced their emergence from hiding in the thickly leaved incense cedars. There could now be discerned a little of the glory of color which was soon to show on the budding trees. The oaks were already rosy with the unfolding of the spring foliage. The newly leaving alders fringing the tiny brook were casting lights of leaden green into the open pools below them.

A slight movement and a faint crackling noise at the stream edge caused me to stop and look inquiringly into the tangle of green brush. Under the protective cover of leafage I saw a mule deer hiding. Soon he came forward, stood gloriously in full view, and began to drink. Then four others came up from behind, thrust

DENIZENS OF THE MOUNTAINS

their way past him, leaped the stream, and came onto the meadow. They saw me but seemed little afraid.

FEMALE MULE DEER

Finally one of them, a large male adult, approached, instinctively thrust forward his nose, and stopped within about twelve feet. Then standing alert with the

THE MOUNTAIN WOOD RAT

large ears fully erect he listened, sniffed the air, and soon gave sign that he was uneasy. With fine control he wheeled, stood still a moment, and then cautiously retired into the brush. The other deer, remaining all the while close together and yet at some distance, paid no attention but went on grazing and browsing. They were in playful mood and manifested their high spirits in banterings and frolicsome chases. Full twenty minutes they remained before me and then, as silently as they had come, they withdrew into the sequestered shelter of their forest home. By leisurely following them I had them within sight full twenty minutes more.

An experience such as this was just what I needed to put me in key for the vigorous climb I had determined to make that day to the crest of the range. As later I worked my way up over the rocky trail the peaceful picture was ever before me, lending enchantment to every scene.

My course took me up hill and down through pine forests of solemn green, along water courses bordered by clustered birches and tangles of aspen full of sweet woodsy odors.

Long before noon clouds ominous of rain began to appear in the western skies. Hurriedly eating my lunch of brown bread and dates I pushed on, winding up my journey late in the afternoon in a small mountain meadow near timber line where I pitched my little tent in a cozy willow nook. The sky was now overcast with

DENIZENS OF THE MOUNTAINS

leaden clouds, and gray showers began to fall. It was necessary to have all my saddles and pack boxes in the shelter with me, and it was with difficulty that I arranged a place for my sleeping bag for the night. Encumbered with the weariness of my journey, I retired early and slept soundly.

About midnight I was awakened by feeling something pulling and twisting and tugging with might and main at the hair on the crown of my head. Quickly I thrust back my hand and was just in time to touch some escaping animal. What it was I was wholly unaware, until I turned on a flash light and saw a much surprised and frightened squirrel-like creature standing on the pack boxes and looking rather indignantly at me for having disturbed him in his labors. It was a wood rat with big brilliant eyes, large rounded ears, and a whiskered face full of animated expression. By what lively fit of imagination did he conclude that I would lie there and let him have what he wanted of my hair for a lining of his nest I do not know. I frightened him off, but no sooner had I put out the light than he was back again, this time to gnaw at the pack boxes that he might get at the food inside. Though I repeatedly disturbed him, he as persistently returned, and almost the whole of the night's sleep was lost. In the morning when daylight came I began to look for his living quarters and found them in a cleft of rocks just behind my tent. I no longer wondered at the cause of his over-

THE MOUNTAIN WOOD RAT

familiarity. I had, as it were, moved right into his front yard.

The morning's wandering early convinced me that I was in an area tenanted by many wood rats besides this familiar neighbor of the night. To locate a wood rat dwelling is an easy thing, for about the rock piles that they occupy there are always small piles of short sticks and the characteristic excreta lying plentifully near the entrance. The surest indication of occupancy is a few fresh green twigs or freshly cut herbs about the entrance of the nest.

A second night spent in the infested area put me to further inconvenience and trial, for now I was not only distrubed by constant visits from the industrious neotomas but called upon to witness the destruction of my boots and pack saddles. While I slept, the mischievous creatures nibbled holes in my footwear, gnawed the laces into shreds, and thoroughly cut to pieces the straps of the pack outfits. Such annoying confidences are not conducive to patience, and after making such repairs as I was able I moved on to new camps where I might have better neighbors.

When mountain wood rats establish their living quarters in mountain cabins, the spaces between the walls and above the ceiling are soon littered with all sorts of rubbish, sticks, stones, old bones, egg shells and can tops. Many persons seeing these things have supposed that the pack rats are actually given to bartering,

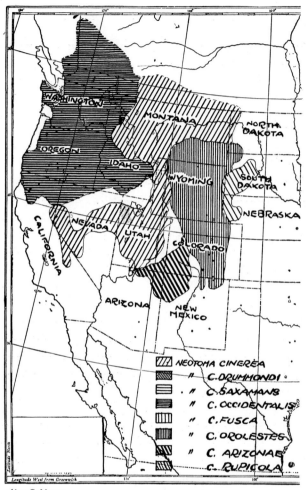

After Goldman

MAP SHOWING DISTRIBUTION OF SUBSPECIES OF BUSHY-TAILED WOOD RATS

THE MOUNTAIN WOOD RAT

never taking away anything without leaving something in its place. And there are those persons of exaggerated imagination and agile tongue who have set agoing the myth that the trade rats even secretly enjoy making a good bargain by leaving worthless pick-ups in pay for our more valued treasures.

The wood rats are often correctly accused of making most of those thumping and galloping noises that so often frighten new dwellers in mountain cabins and keep even old timers awake half the night. These animals are so extraordinarily industrious at nocturnal noise-making that a single pair of them in the attic of a house can furnish enough racket to set anyone to thinking that two dozen hobgoblins are about. They run over floors of the attic and climb the rafters. They thump on boards with their rear feet for long periods at a time. They drag sticks and all sorts of trash and trinkets between walls and on the top side of the ceiling overhead. And if there is any additional way within their capabilities to make themselves busy and noisy it seems they are sure to find it.

The reputation of wood rats as racket makers was early established, for in one of the first records in American literature, in which a wood rat is mentioned (Khan's Travels, English edition, II, pp. 47 and 48, 1749) it is spoken of as a rock dwelling animal occurring in great numbers and "making a terrible noise."

The range of food relished by the bushy-tailed moun-

DENIZENS OF THE MOUNTAINS

tain wood rats is considerable. Green food is much eaten as are also the seeds and outer bark of certain trees and shrubs. When camp scraps are available, these are generally taken in preference to other natural foods. Edward R. Warren has in his "The Small Mammals of Colorado," published by the Colorado Mountain Club, a photograph of a wood rat eating miner's candles. They will even eat soap, but we must not be surprised at this, for the round-tailed species of wood rat which inhabits the deserts is at times very fond of it, as are also certain of the wild mice.

The most common faunal associates of the mountain neotomas are the conies which share with them in security, and we hope in amity, the crevices of the rock slides.

CHAPTER II

THE CONY, DENIZEN OF THE ROCK SLIDES

AT THE base of nearly all the great and noble cliffs which abound in the Rockies and the High Sierras of California are found enormous rock accumulations known as talus slopes. These are the ruins of the cliffs themselves, and like giant buttresses slope up to the sheer rock faces as steeply as the material will lie. Generally these accumulations of rock debris are so devoid of soil or humus that little plant life exists upon them. Especially is this true if they occur at high altitudes near tree line, where the cold winds sweep over them with such inhospitable furor during a major portion of the year. Bleakness then reigns supreme. A few grasses and alpine herbs and occasionally a stunted tree may eke out a miserable existence, but they can do no more. It seems about the most unlikely environment one could imagine for any animal to choose as its abode. Yet this is the very one selected by that midget mountaineer, the cony or little chief hare.

Our North American conies are thick-set little mammals and look for all the world like tiny, round-eared, tailless rabbits. They are about seven inches long from tip of nose to base of tail, but they sit in such hunched up positions that to the observer they seem scarcely

IN THE DEBRIS AT THE BASES OF THE CLIFFS THE CONIES LIVE

THE CONY

longer than four. Their facial expression is very remarkable and is such as to impress one that they are very bright, active and circumspect little creatures.

SIERRA CONY

One of the most pleasing experiences I have had with these denizens of the rock slides was just below Sawmill Pass in Inyo County, California. This outlet to the

DENIZENS OF THE MOUNTAINS

desert slopes of the Sierras is a broad saddle with an elevation of more than eleven thousand feet. My companions and I had descended about two hundred feet below the summit and camped in a small rocky amphitheater the floor of which was well covered with grasses and lovely alpine flowers. Snow banks were on its sides in many places and from their bases trickled a number of tiny streams. No sooner had we unpacked the burros than we discovered we were in the midst of a colony of conies. It was after sundown but they were yet out and active for we heard plainly their squealy voices, the characteristic *eeh—eeh—eeh*, rapidly and sharply aspirated, and also another single somewhat hawklike note. These calls were continued more or less frequently all through the night, and there was confirmed to me anew the statement of an old ranger that the conies sometimes chatter long after dark, especially on moonlight nights.

Very early in the morning, long before sunrise, I heard one of my conies just above camp. By carefully scanning the rocks I located him sitting hunched up on the point of a large angular piece of granite where he was greeting the day with his innocent barking. Being eager to find his retreat, I quickly dressed and with my eye still upon him crept upward among the stones. He uttered his sharp cry of alarm *eeh—eeh—eeh* again, and then quickly and defiantly retreated into the crevices of his rocky home. As I came to the place where he had

THE CONY

been sitting I heard short subdued squeaky notes issue out from under the talus blocks, and it seemed easy to imagine that the rocks themselves were alive and speaking to me. A note came first from one crevice and then from another, as though the cony was scurrying about in the galleries under my feet, occasionally taking peeks and throwing challenges to me through the chinks in the rocks. After walking about a little I came upon the den. Its location was indicated by accumulations of shotlike excreta and by a small stack of "cony hay" beneath a shelving rock. During the sunny days of the short alpine summer this little harvester had busied himself curing plants against the days of winter hunger. Not only did he have grasses out drying, but also sprigs of gooseberry, and pieces of heather and limber pine to vary his winter fare. The last plant he had carried from a tree fully one hundred and fifty feet away. Some of the material was yet fresh but most of the hay was already well cured. There must have been a pailful of it. Other little stacks of dried plants which were nearby appeared much weathered as though they had been left from a former season. A number of white columbines grew but a few feet away and almost every bud and blossom had been plucked by the industrious little harvester. Alpine heather and grass predominated among the material that had recently been gathered. The grass was cut long (ten inches or longer), but the twiggy vegetation was taken in lengths not greater

DENIZENS OF THE MOUNTAINS

than four inches. As I wandered leisurely about inspecting the vicinity of the homesite I found many pieces of heather which had been cut and carelessly dropped along the way; probably they had been forgotten by the cony in fits of absent-mindedness.

Just above the pool where I obtained my water was a slide of very large rocks. A cony had established his home there and when he heard me splashing in the water his inquisitiveness was so aroused that he came out and perched on an angular granite slab where he could observe me to advantage. Having satisfied his curiosity and pronounced me harmless, he went again in among the crevices, but not for long. I soon found that I could draw him out almost at will by imitating his bleat. The giving of this call note acted almost like magic. No sooner did I sound it than he popped his head into daylight, and often he vociferously answered back as if to reproach me for my rude fun. It was hugely amusing to note the appearance of intense earnestness that characterized his movements when he did this. The cheeks were contracted and the ears twitched; indeed the whole body was so vigorously jerked forward that there could be no doubt that he meant every word he said to me.

The conies are quite daring little creatures especially when at the mouth of their den and often retreat defiantly before you. They seem to appreciate the protection their rocky home affords them against large

THE CONY

predators. Except for the weasels they have few natural enemies. Against the attacks of these slender-bodied, bold, and bloodthirsty murderers the conies can offer little resistance, nor do their homes give them any shelter, for the weasels can enter the smallest crevices open to the habitation of conies.

Merritt Cary in his Biological Survey of Colorado tells a pathetic story of a cony which took up its home and built a large grass nest beneath a prospector's cabin located near a rockslide just below timber line. The cony, showing much confidence and neighborliness, appeared often in the cabin, coming up through a broken board in the floor. Finally a day came when it did not make its usual appearance but in its stead a dwarf weasel was later seen at the hole in the broken board, peering in all directions and craning its long neck with all that bold curiosity characteristic of its kind. Fearing for the welfare of the cony the prospector killed the tiny cut-throat, but apparently too late, as he saw no more of his interesting and friendly companion.

There is always a great commotion among the members of a cony colony when a bird of prey hovers over it. Immediately sharp notes of warning arise from members all over the colony and there is then a grand scrambling for shelter on the part of every one exposed to the gazing eye of the enemy. The colors of the conies' thick and fluffy hair must prove a protection of no mean

DENIZENS OF THE MOUNTAINS

import to them, for it blends very closely with the grays and browns of the rocks on which they dwell. Time and again I have noted the near impossibility of spotting one of the tiny animals when it was sitting still on slide rock.

In spite of the fact that conies are dwellers of high mountain regions, where snow and cold necessarily prevail during more than half the year, they do not hibernate. The interstices between the deeper granite blocks of the talus slopes and moraines are quite free from snow throughout the winter, and in these crannies the conies find ample shelter and live in snug comfort during the coldest days.

Three or four young are born during the middle part of summer. They are said to be as precocious as young jack rabbits and to begin foraging independently when one-fourth to one-third grown.

CHAPTER III

THE SAN BERNARDINO CHIPMUNK

WHO is not pleased with the company of chipmunks in the forest! Contentious little beasts they are, nosing into everybody's business, restless, inquisitive, and ransacking every nook and crevice for food and opportunity for mischief-making, yet with all these apparent faults the merriest, most pleasing midgets of the forest. Indeed to most of us, mountains could never be "real" mountains without them. The more plentifully they are about, the better we like it. None of the animals seems to get more pleasure out of life and none is able to impart more of it to their human companions. From early morning until almost dusk they may be seen spending their seemingly "happy-do-nothing-days" energetically scampering over the logs and running around in piles of rock or rotting bark. The sharp-pointed and erect ears, the tapering striped muzzle, and brilliant jet eyes give these quick restless and wary little creatures a most knowing expression, and when one sees in all those movements of the bushy tail and in the earnest chipperings the expressions of an active little brain, he is more sure than ever that he can not be content to be unfamiliar with their ways.

Chipmunks of the genus *Eutamias*, to which all of

DENIZENS OF THE MOUNTAINS

our true western chipmunks belong, are very widely distributed in the mountainous regions of western America. Indeed, one may have the company of one or more of the many species in traveling from southern Alaska to middle Lower California. So many species and subspecies of this animal are listed by specialists in mammal study that the average traveler will come quite to despair in an attempt to individualize them. (California lists nineteen; Utah, eight; Colorado, eight; and many of the other western states as many as three.) Since their mental make-up and habits are quite similar, an acquaintance with several of the more widely distributed forms will put one in possession of a very good knowledge of chipmunk ways.

All of the species dwell in well treed or shrubby districts where acorns, pine nuts and other seeds afford them ample food. If the traveler thinks he has seen chipmunks in the barren deserts he may rest assured that he has mistaken one of the small desert-dwelling ground squirrels for a chipmunk.

There are several kinds of chipmunks, such as the Gila chipmunk of Arizona and the sage-brush chipmunk of California and Nevada that inhabit broken rocky country almost entirely destitute of trees, and where only low shrubs predominate. Most of the species, however, are attached to the coniferous forests. Each species has its own definite range and to understand fully any one of them we must thoroughly acquaint

THE SAN BERNARDINO CHIPMUNK

ourselves with its environment; for as Mr. Sclater has observed, "the habitat of an animal is as much a part of its definition as its structure or external form."

SAN BERNARDINO CHIPMUNK

In the higher coniferous forests of southern California lives the San Bernardino chipmunk (*Eutamias speciosus*), a species which for many years I have watched with the greatest interest. In general its habits may

CHINQUAPIN THICKET: HOME OF THE CHIPMUNK

THE SAN BERNARDINO CHIPMUNK

be taken as typical of most of the high, tree-frequenting forms. It is most abundant in the chinquapin thickets of the upper part of the yellow pine belt. This chipmunk is exceedingly fond of the nut kernels found within the chestnutlike burrs of this plant, and it finds beneath the twiggy branches ready cover for hiding when the approach of an enemy demands a quick retreat. It is almost equally partial to rocky areas in the forest where there is an abundance of fallen timber. I have repeatedly noticed how chipmunks, when possible, choose as highways the prostrate trees and barren rocks. Almost invariably they will, when surprised, take to a log and run its length rather than race along to safety over ground. If a standing tree offers the nearest and only haven, they can climb it with a dexterity that must seem remarkable even to a steeple jack.

Anyone who examines a chipmunk in the hand will be struck with the amazing similarity that exists between its foot and the foot of a bird. The little slender digits are much like the toes of birds and are armed with long, sharp clawlike nails. The possession of these good grapplers often means to the chipmunk the preservation of life. The animal is so small that he can show no front to his larger enemies and the best he can do is to make a rapid exit on foot from the scene of danger. If you have ever seen a frightened chipmunk, you know how well he makes his hurried, sputtering retreat up the tree. Generally he manages to go up on the side

DENIZENS OF THE MOUNTAINS

of the tree opposite you, not stopping until he is well up among the branches. Then, mounting one of the side limbs, he proceeds in stillness to watch you, peering curiously from out those fine, lustrous, black, beady eyes, with all the inquisitive expectancy at his command. Feeling his security, he may now proceed to scold you, giving his quick, saucy, defiant note with a regularity that seems almost machinelike in its timing. If you sit down under his tree he will look down upon you for ridiculously long periods, almost wearing out your patience and your eyes also.

I recently came upon a little fellow feeding in a young silver fir. He was so upset with excitement at my approach that in his effort to get away he ran distractedly first out to the end of one branch and then in rapid succession to another and another. Several times he almost lost his foothold. I sat down and tried to remain perperfectly motionless but he was not to be induced into thinking his perch was longer a place of security. Slowly and cautiously he crept down the farther side of the main trunk of the tree, then dashed groundward and across a small clearing to a rock pile where he could secrete himself. But curiosity soon got the better of him and out popped his head to see whether I was still on his trail. Thinking to make his situation still safer he now skimmed along a log to a crevice in its further end. It was indeed amusing to note how, even in flight, the little fellow found time to make himself

THE SAN BERNARDINO CHIPMUNK

comfortable. Stopping abruptly, he went after an annoying flea with a bewildering rapidity of scratching movements, and then in a fraction of a minute was running again.

These midget mountaineers are somewhat active in winter as well as in summer. Hibernation in so far as I can learn is partial or irregular. Those living near timberline or lower where both days and nights are constantly cold, are probably never out in winter. Those living in the yellow pine belt emerge at times during sunny weather and may be seen nosing under leaves on the patches bare of snow, where perchance they may find some titbit left there during the summer days of plenty.

The time when the young are born is greatly dependent on the altitude but is probably always during late June and July. There are four or five to the litter and these when born weigh but a fraction of an ounce. The nest is made under stones or in hollows of trees and lined with soft vegetable fibers. The juveniles in their fuzzy new bright colored coats are extraordinarily attractive little creatures, and since they are without vestige of fear, they become unusually familiar if coaxed to camp by the offering of food. At this time they are so easily tamed that many a little fellow pays a penalty for his innocent ways by being forced to spend his future days as the captive of man. The only drawback these young pets offer is, as Robert Rockwell

DENIZENS OF THE MOUNTAINS

says, their tendency to "become too tame," in which event one never knows whether they are busy in the flour bin or asleep in one's coat pocket.

The strongest bond that attaches the young to the parent is, I believe, hunger. Whenever I have seen very young chipmunks about the nest I have never been able to see much solicitude manifested for them by the parent. But in spite of this parental coldness, the little family circles are often not broken up until late in the summer.

The play activities of the young chipmunks about the home log are most varied, and serve as an endless source of enjoyment to the patient observer. Often I have seen one sitting on a stump when another approached him; the two touched noses and then with a frolicsome leap, the first jumped "clean" over the other. The little playmates, turning, were soon presenting faces, exchanging nose greetings and making themselves ready to repeat the performance again. It is a curious sort of leap-frog game common enough about chipmunk nurseries. Repeatedly I have seen two young chipmunks approach each other from opposite ends of a log, present faces and then with jerky motions wheel round and round, merry-go-round fashion. Playing tag, however, seems to be the popular game, but I can hardly say the favorite delight of the adults, for quarreling is with them an almost constant pastime. Indeed, they spend so much time quarreling that it

THE SAN BERNARDINO CHIPMUNK

seems, at times at least, there would be little chance for them to engage in the more serious work of providing food and shelter for the coming winter days. But never you mind, there is a serious side of life and the winter needs are taken care of well enough.

PANCAKES FOR BREAKFAST

If the camper happens to locate in a district "infested" by chipmunks, he may find their familiarity almost a menace to his patience and good temper. He may leave nothing edible exposed for the birds or his domestic animals that these clever and persistent ro-

DENIZENS OF THE MOUNTAINS

dents will not get. Yesterday I saw a chipmunk carry off, not once but twice, whole pancakes and even try to take one up a tree. Their internal cheek pouches which serve them "as substitutes for suit case and hunting coat pouches" are very capacious and, if undisturbed while feeding, chipmunks will stuff them so amazingly full that the little chaps appear, as one has aptly said, to have an aggravated case of double toothache. Doctor Mearns tells us how the Arizona chipmunks (*Eutamias cinereicollis*) in their bold quest for food even climbed into the nose bags of the horses for grain, frightening some of the animals into breaking their halters.

Early one September I chose as the site for my camp the edge of a little meadow where I could enjoy watching the coming of the autumn colors on the leaves of the shrubs. In a mat of buckthorn just across a little gulley was a tall upright stump of a dead conifer near the top of which were several holes resembling those made by nesting woodpeckers. Into one of the cavities I repeatedly noticed a chipmunk disappearing with something in his mouth and later reappearing at the entrance, his mouth empty. Always at about the hour of sunset he seemed particularly active. With an inquisitiveness for which I afterwards chided myself, I one day set to opening the hole with my axe. The hollow was not more than two feet deep and as I came to the bottom of it I saw what it was that served as the

THE SAN BERNARDINO CHIPMUNK

chipmunk's attraction. This was his home and there at the bottom was the snug little nest he had been making to keep him cozy and warm during the winter days to come. Of finely shredded bark of wild lilac and plum had he made it. I disturbed it as little as possible and with a nail replaced the slab of wood and bark I had torn away, hoping that the little hermit of the woods would find his cell still tenable and good for the purpose he made it. I found no store of provisions, but that was not surprising, for these small rodents are more or less active throughout the winter, or again, it is quite possible that he had his food supplies cached elsewhere.

Chipmunks eat such a variety of food, both animal and vegetable, that they stand a much better chance of making ends meet through the winter than animals of a restricted diet but which store more food. If there is any time for lean feeding it is probably in the spring when the winter stores are depleted and before either the insects are abundantly out or the plants have begun to grow again. As soon as the nesting season for the birds begins, the chipmunk's hunger worries are over. There is then active plundering of the nests of the smaller species of birds for eggs. Indeed such pilferers of eggs are the chipmunks that if many of the birds did not keep almost continual vigil over their nests there would be few young to rear. But perhaps the birds are partly to blame. So often do they boldly

DENIZENS OF THE MOUNTAINS

and ridiculously expose their eggs in flat and open nests to the gaze of every comer that we can not but wonder that the chipmunks on their daily rounds of discovery do not find and rob more of them.

All of the chipmunks are very fond of grass seeds in the milk. As the panicles become heavy with their loads of ripening grain, these little harvesters feed almost exclusively on grass seed. To secure and bite off the seeds, they generally sit erect and hold the stem between the fore paws. One recent afternoon, a chipmunk came out on a rock not more than four feet from me and showed me his clever method of getting grass seed from unusually high stems. He first stood upright on his haunches, then grasping the stem near its base with his fore paws he bore down on it hand over hand, in a most human way, until the seed came to the level of his mouth. Sometimes the seed is eaten on the spot but as often it is snipped off and taken to an exposed place on the rocks where a better outlook is afforded. A chipmunk when feeding can never afford to be off guard. The "rule" among them apparently is: Keep out in the open to see what is going on. If anything suspicious comes near don't wait to find out what it is but flee for cover. Afterwards an inquisitive reconnoitering may help you to ascertain what it was that frightened you. Having satisfied yourself that no danger lurks near, you may get out into the open again and resume your feeding activities.

THE SAN BERNARDINO CHIPMUNK

Two kinds of notes are uttered by this chipmunk of our sketch: One is a regularly recurring *pit—pit—pit*, given when the animal is reconnoitering about on the rocks and trees. It is evidently the chipmunk's sign of concern. The other note is one of alarm, a sputtering series of chippers rapidly given. It is generally heard when the chipmunk is running for cover along a log, with his tail held straight up in the air.

Recently the little penetrating *pit—pit* note, repeatedly given for an extraordinary length of time, made me curious to know its cause, and when I followed it up I came upon two chipmunks in a Jeffrey pine. One of them was eating the small staminate pollen-bearing cones on the end of a twig and the other was seated on a limb flashing his tail from side to side and doing all the talking. Down below on the ground was a gopher working and throwing up huge armfuls of earth. The little chipmunk was intently peering down at him, filled both with curiosity and concern at the gopher's movements. The other chipmunk was at first not at all disturbed over all this, but went on eating his small "cones," stripping them one after another after the manner of a tree squirrel. But after his companion had spent some ten minutes "barking" at the gopher, he dutifully joined him, and they now both told the busy rodent just what they thought of him. With the earnest actions of these little scolders a comic element came in, for the mute old gopher worked on as unconcernedly as though they had never existed. By his indifference I could hear him saying: "Next time, little folks, you give tongue to hot words choose someone to scold who thinks you more important and who will be more terrified by your remarks."

CHAPTER IV

THE MERRIAM CHIPMUNK

FOR many successive summers I have sought a retreat far back in a secluded portion of the San Jacinto Mountains where a merry little mountain stream runs its short course through a narrow but shallow valley and then tumbles over a ledge to make its precipitous drop down the steep slope of the mountain's north rim. It is a quiet place, protected from the harsh winds. The trees have grown symmetrical, strong, and tall, and dainty stemmed flowers and grasses form a fine forest carpet everywhere beneath them. A hundred yards from the flower-bordered stream, the mountain rim, which runs parallel to it, falls away downwards thousands of feet and offers a great, open, inspiring view of the wide reaches of the Colorado Desert beyond—a prospect so sublime that every faculty of the soul is involved in the full comprehension of its quiet and awful beauty. To this place of good view I often repair in the morning at daybreak to watch the sunrise. There in the evening, at the hour of sunset, I see the great blue shadows of the mountain creep eastward over the silent, yellow sands of the desert floor and on across the sun-baked, rock-ribbed arid mountains beyond. Silence is there such as I have found in few places,

DENIZENS OF THE MOUNTAINS

stillness so intense that the whole soul is left untrammelled and given entire freedom for the comprehension of beauty. The Place of Soul Spur I call it, and as often as I seek it I am fed with inspiration.

One particular evening as I was peacefully lying on a great granite slab watching the magnificent flight of a golden eagle which was gliding about with eight thousand feet of rare air below him, my attention was suddenly diverted by a little rustling sound at my feet. As I glanced down I saw on the edge of my rock, in alert and expectant attitude, a tiny baby chipmunk. I could not have wished for a more unusual and fitting place to meet so choice a friend. I remained perfectly quiet and for many minutes it played about me without suspicion or fear. Just below me on the desert slope of the mountain was an enormous granite buttress with vertical sides of a hundred feet or more, and on this I saw another chipmunk, perhaps of the same family, running upward fleet as a lizard. Nor did it stop until it reached the top of the cliff where a gnarled and scaly mountain mahogany grappled for hold in a crevice of the rock. There it sat watching me, conspicuously handsome in its new striped coat. Inspection of its color markings soon revealed it to be a Merriam chipmunk.

The finding of the small mammal here at this altitude was such a surprise to me that I proclaimed my discovery aloud and unwittingly scared all my little neighbors into the rocks. The place is one of the

THE MERRIAM CHIPMUNK

highest recorded stations for this chipmunk. The altitude here is almost 9000 feet, while ordinarily 5000 to 6000 feet is the upper reach of its domicile. The explanation for this high station is found in the occurrence of a cover of low shrubs and an arid climate brought about by the dry winds which sweep upward from the desert below. The locality became additionally interesting and unique to me when later I found the San Bernardino chipmunk here intimately mingling with the Merriam. Since then I have frequently noted the two species feeding in the same area and apparently on good terms, the one keeping pretty well to the timber while the other stayed on the ground in the brush tangles below. When the two species are found together it is easy to distinguish the larger Merriam chipmunk from his small tree-dwelling cousin.

To the north of the Colorado Desert is a range of most arid, rocky, and forbidding mountains, marking the dividing line between the Colorado and Mohave deserts. Few persons would suspect that on their north slopes they carry a goodly growth of piñons and junipers, and that here is found one of the most interesting floral and faunal environments in the southwest. The altitude is between 4000 and 5000 feet. Wild rocky gorges and peculiar domes of disintegrated granite predominate the landscape, and it was a surprise to me to note in a place so far remote in the desert ranges the presence of the Merriam chipmunk. Indeed

DENIZENS OF THE MOUNTAINS

it marks the most southeasterly extension of its range. Here it was found gayly scampering about on the rocks among the oaks and piñon trees.

Rains are very infrequent in this region. Sometimes a year elapses between storms when water falls in quantity. Cloudbursts bring most of the moisture, and the cattlemen have found it necessary to build dams to impound water sufficient for the needs of their stock. At one of the larger reservoirs I found that a family of Merriam chipmunks had taken up quarters in a rock pile within two feet of the water's edge—a rather insecure, perilous site for a home it appeared to me, for a cloudburst might at any time raise the waters and submerge it. But the uncreated future is with these small animals unthought of; conscious life is concerned only with the actual present.

The waters of the little lake had attracted a great number of birds, for it was one of the few places for miles around where they might procure a drink. The high rocky walls were frequented by many rock and canyon wrens, and the air was fairly alive with noisy Cassin kingbirds. At this particular time a number of ducks were there too, wheeling about in the air above, too much disturbed to settle down for long while strangers were near. A white-faced ibis was wading in the mud flats nearby, often rising in air and with deliberate wing motions circling above the chipmunk's home. But to all this commotion of birds, this modest

THE MERRIAM CHIPMUNK

little family paid slight attention. One of the kingbirds hawked a fly within two feet of them, yet the chipmunks scarcely moved a head. But when *I* came up they all ducked for cover! After I had remained in perfect quiet for a time, a little, self reliant, half-grown chipmunk ventured up close to me, settled himself on a rock, and, with his head well down, in most business-like manner spent a full fifteen minutes "barking" at me. He quit only when one of the parents came up and chased him off. One would imagine the adult saying: "You foolish little man! Don't you know better than to risk your life that way? That's a pretty big animal you're talking to."

The tag playing instinct was well exhibited in this chipmunk family and a lot of time was whiled away by them day after day in chasing one another about on the black rocks. The explosive chippers and soft chuckering notes were most interesting to hear and it was pleasant to imagine the "conversations" these little people of the wild carried on among themselves concerning the things that happened in their small world beside the lake. I noted amusedly how their "chatters" extended to the very end of the tail, and how each outbursting note brought with it a decided jerk of that fluffy appendage. The tail in this species is generally carried well up, and in emotional states the chipmunks whisk it about much as cats do when excited on the watch for game.

When these observations were begun it was in the

spring of the year (early May) and the chipmunks were still wearing their dark winter coats. Because the hair of the adults was then badly worn, the basal dark leadish or blackish portion of it was exposed. So dark were these specimens that at a short distance they appeared wholly black, *without sign of stripes*. The youngsters were dark also, a color which they held until after a post-juvenile molt, at which time they assumed the bright summer pelage. The attractive colors carried by the adults from June to September are assumed only after the breeding season.

Since the Merriam chipmunk is essentially brush-inhabiting and occurs in an environment best suited to the life of the more aggressive mountain-dwelling reptiles, the gopher snakes, garter and rattlesnakes, it finds in these reptiles some of its worst enemies. On my mountain travels I have killed a fair number of rattlesnakes and a surprising proportion of them contained the remains of Merriam chipmunks. The rapacious Cooper hawk, whose most urgent business seems always to be the satisfaction of its hunger, must many times spy them out while they are disporting themselves on the rocks and take toll of prey from among their numbers. Except for the occasional murderous-minded boy, these sprightly rodents seldom find in man a serious enemy. Hunters consider it beneath their dignity to shoot such tiny creatures for sport and hold their flesh unfit for food.

THE MERRIAM CHIPMUNK

Generally the chipmunk home is made in a hollow tree or stump or under a fallen log; once in a while a deserted trade rat's nest is occupied. All such situations are unfortunate retreats when the devastating brush fires sweep over the mountains, for if you have observed at all closely, you must have noticed how clean the hungry fires sweep every burnable thing in their path. Chipmunks, with practically all the other small forest denizens, perish miserably in the flames.

Four or five young are born in the last half of May or in June, the time varying with the climatic conditions. Individuals living upon the low mountains within the desert area breed and bring forth the young much earlier; of this I am sure for I have seen one-third grown young the second week in May.

Living in the dry brush environment where the heat in summer is often intense at noonday, the chipmunks seek water when they can get it. Springs are visited, and after rains shallow water pockets on the tops of rocks give them opportunity to drink. Other sources of water are found in wild berries, such as currants and gooseberries. They feed on a variety of food including acorns, juniper berries, piñon nuts, insects, mushrooms and many kinds of seeds of grass and other plants.

While the chipmunks inhabiting the brushy areas of the California mountains have, because of their activities and abundance, to a certain extent made necessary the abandonment of direct seeding work on the part of the United States Forest Service, I have the word of E. R. Munns that they are often of great assistance in bringing about reforestation of denuded areas. "Many areas," says Mr. Munns, "which normally would be without timber have been seeded by the action of the chipmunk who buries small pockets of seed during periods of heavy seed abundance, and my own feeling is that on many of our forest areas the presence of reproduction is due entirely to the action of this very busy little rodent."

The Merriam chipmunk was named in honor of Dr. C. Hart Merriam, founder of the U. S. Biological Survey and long a leading student of mammals of North America. The late J. A. Allen in his monograph on the chipmunks in 1890 said, "It is safe to say that seven-eighths of the really good material forming the basis of the present paper owes its origin directly or indirectly to Doctor Merriam's influence."

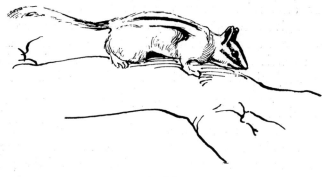

CHAPTER V

THE MOUNTAIN CHICKADEES

WHEN all the summer bird visitants have left the mountains to seek more genial winter climates, the hardy chickadees remain to enliven the woods and gladden us with their merry chatter. On cold and leaden-skied days they make ones walk almost as promising of pleasure as any that may be taken when the sun is shining. As we see them sallying forth in merry little troops while lisping notes of gladness and blithely and nimbly bustling about from limb to limb, we catch a certain spirit of merriment with them. We can not but feel pleased, after all, that winter days are with us again.

The mountain chickadee makes it his business to sing almost every day of the year. Only during the molting time do his cheery notes lag. He has two songs which stand out distinctly unique and prominent. The first is a fine merry whistling *deedledee-dee-dee-dee*. The second is a series of four plaintive high-pitched minor notes, particularly conspicuous in the spring of the year but uttered more or less at all seasons. You can never be certain as to how many of the series of notes of either song is to be rendered, for sometimes the winsome little singer capriciously stops right in the middle of his ditty or begins in the heart of it and utters but

DENIZENS OF THE MOUNTAINS

the last two notes. All sorts of variations may be noticed, and when one comes to making record of the chickadee's complete repertoire he finds it a very full one. Indeed one can never be quite certain as to when he will not hear a new note or a different combination or inflection. This is especially true when the chickadees are in company with others of their kind or are consorting with their near stubby-tailed neighbors, the nuthatches. One can not but feel that perhaps after all these birds have a language that is full of meaning though to us it may seem quite unintelligible.

A chickadee's motions are as varied as his blithe and lively songs. His powers of flight are inferior to those of many of his strong-winged relatives, but few of them can match him as an acrobat. Hardly is he still a minute. See him there in that silver fir clinging upside down on the branches. Now he goes wheeling round and round, then up he flies to the trunk of a pine, only to descend it head foremost. Soon in the abandonment of light-heartedness he dashes madly away to another tree and repeats all over again his funny performances. Some of the maneuvers seem to be made in birdish play but most of them, we feel certain, are the accompaniments of his business pursuits. He earnestly concerns himself with the taking of insects that are on the under side of the limbs as well as those on top, and many a twist and turn about the branches he finds necessary to complete the inspection.

THE MOUNTAIN CHICKADEES

While most of the chickadees stay in the pine forests during the winter season and brave the snow, sleet, and blustery winds, some individuals descend the mountain slopes and spend their winter days about the cottonwoods and mesquites of the cismontane valleys. Often during storms in the mountains I have seen them feeding far out on the desert floor with the intermediate sparrows, but I have found that they always go back toward their mountain home as soon as the weather moderates.

We are naturally led to ask how such little bundles of flesh and feathers are able to brave the cold of their high mountain habitats so well. It must be borne in mind that a bird possesses in his body a remarkable power and heating plant. The body temperature is kept unusually high by a very rapid oxidation of the food stuffs consumed. Further, the heat which is generated is kept within the body by a coat of fluffy feathers, the best heat-retaining covering known.

The most common chickadee of the Sierra Nevadas and Rockies is the Gambel mountain chickadee (*Penthestes gambeli*), a very near relative to the Bailey chickadee of the southern California mountains, and approximating closely the latter in its habits. The Gambel chickadee is closely associated in the Rockies with the long-tailed species (*Penthestes atricapillus septentrionalis*), a bird much resembling the common eastern chickadee but paler in coloration.

DENIZENS OF THE MOUNTAINS

Dennis Gale, an old time naturalist of the Colorado Rockies, tells us so engagingly of the nesting habits of the little Gambel chickadee that I can not refrain from transcribing from his note books (now preserved in the University of Colorado Library) his description.

"The favorite choice for a nest site," says Mr. Gale, "seems to be abandoned cavities made by the red-naped sapsucker (*Sphyrapicus varius nuchalis*) by far the most industrious and capable of all our woodpeckers for hole making—who having pierced the hard outside zone of living wood of the aspen and well advanced the larger excavation in the central part will often abandon it half executed, feeling perhaps that it is not suitable for its requirements. But it is quite sufficient for little *Parus*.* This home selected, the little female holds against all comers; never leaves it except for a moment, perhaps from the time she makes her nest till the young ones are quite well grown and their demand for food requires the joint labors of the old birds to furnish it. The male is very affectionate and attentive, feeds and waits upon the female during the period of incubation and while the brood is yet young, and seemingly is ever within call. The devotion of the female to her charge is very remarkable, sacrificing her life if necessary before she can be induced to desert it. You can not frighten her off her nest by hammering the outside of the tree ever so hard. In one instance upon chopping

*The generic name formerly applied to the chickadees.

THE MOUNTAIN CHICKADEES

the cavity, it was half filled with chips in the operation. Removing them by hand, what was my surprise to find her buried with them, and still covering her eggs! I had to lift her off, and rejoiced to find she was unharmed. Such devotion affected me deeply, I am not ashamed to say."

CHAPTER VI

THE ANTHONY GRAY SQUIRREL

HAVING taken the pack boxes from my donkeys that they might rest and eat during the noon hour, I led them back over the rocky trail to a place where I had before noticed a small grassy plot. This led me under some giant yellow pines where I soon saw an exhibition of animal activity so ludicrous and unique that it stands out among the most interesting of my experiences in the field.

Hearing the wild crying of a number of California woodpeckers and a rattling of claws on the bark of a tall dead tree, I looked to see the cause of all the commotion and noticed a fear-stricken gray squirrel desperately fleeing up a tree trunk, mobbed by half a dozen woodpeckers who were wheeling about and dashing toward him in uncontrolled and furious anger. "An egg pilferer being brought to justice in the court of the woods," I cried, while I noticed other woodpeckers flocking in to assist in this affair of mutual interest. At loss for a better plan of action, the frightened, fugitive squirrel was whisking around and around the tree, evidently hoping that by always going to the other side he would be free from attack. But the birds were onto his game, and no sooner was he

ANTHONY GRAY SQUIRREL

DENIZENS OF THE MOUNTAINS

at the refuge he sought than his avian pursuers were there to meet him and to take revenge by pulling from his hide copious beakfuls of hair. Again he tried to make good his escape by fleeing obliquely toward the other side of the tree. Again was he met by these birds that were not to be satisfied until full justice had been meted out and sufficient punishment given to fasten the lesson in the pilferer's mind.

After the squirrel had been routed from one position after another until his spiral retreat had brought him to the very tiptop of the tree, he found himself in a most unhappy plight. Why this squirrel went to the tree top for refuge is difficult to conceive, for there, above all other places, was he exposed to the attack of the birds. The instinct to climb above ground for safety probably got the better of his judgment. Tree tops to his mind had always been places of safe retreat. If here he was not safe, where could he go next! To leap to the ground from those dizzy heights probably appeared but a leap to death, and while he was there further losing his wits or deciding what to do next (I am quite unprepared to say which) hair was flying faster and faster, for the birds were giving him no rest. There was indeed now no time to lose, and he finally saw that to descend the tree was the only course left to him. But instead of going rapidly down in a straight course, as good sense would have dictated, he again showed his squirrelish folly by spirally zigzagging his way down, a

THE ANTHONY GRAY SQUIRREL

procedure which proved very expensive if estimated in costs of fur.

When finally this bulging-eyed, bewildered squirrel neared the bottom of the tree, that is, within about fifteen feet from the ground, he came to judgment, ended his foolish and perilous tactics, and plunged downward precipitously through the air, no doubt without thought as to where he would land but glad to get anywhere out of reach of those pursuing birds and out of hearing of their guilt-accusing cries. The landing place was the top of a mat of thorny-twigged buckthorn under the protective cover of which he now promptly fled. My suspicion is strong that that squirrel minded his business well for the rest of that day at least.

A somewhat parallel experience involving gray squirrels and woodpeckers came to my attention in the Palomar Mountains of San Diego County, California, where I have often spent protracted periods in field studies. Above me at my camp was a great live oak of giant spread whose crown stretched upward fully sixty feet or more and covered an area a hundred feet across. Two other similar trees almost as large, and in addition a fine old white fir, interlaced it with their branches to add size to the green canopy. Two young gray squirrels had chosen this beautiful arbor for their home and every morning at daybreak they were up among the branches frisking and capering in high paroxysm of merriment and agility, stopping but now and then to give vent to

DENIZENS OF THE MOUNTAINS

their hilarity and constitutional joy in jocund barking. Transit from tree to tree was freely and easily accomplished, and scarcely were they in one until I saw them scaling the branches of another. Never have I seen more merry squirrels than these, and right good cause they had to be happy in such a place of plenty where acorns or cones were in almost every tree and where they were almost wholly protected from molestation of enemies in the form of down-gazing birds of prey. The only annoyance to their otherwise happy existence seemed to be the activities of the blue-fronted jays and woodpeckers, which, as though they could not possibly see anyone more happy than themselves, frequently flew pell-mell at the squirrels when they were racing most hilariously through the tree tops. I could not quite understand the birds' motives here for this was far past the nesting season, being late in September. Perhaps they suspected that the squirrels might be robbing them of their acorns. But this is only conjecture. Unfortunately animals are unable to speak to us through the use of language, and this most valuable road to a knowledge of their inner conscious life is closed; and as stated by McBride, "No other avenue can lead the inquirer very far."

When at first a group of birds came flying toward them, the squirrels sought the tree tops with greatest haste and then "froze" to a branch, lying as flat as possible. This action seemed to disappoint the birds

THE ANTHONY GRAY SQUIRREL

greatly, for instead of further attacking the squirrels they sat dejectedly by, now and then harshly screaming as though they hoped by this to start them up and so continue the tantalizing fun, if such it was. These squirrels were wise enough to wear out the bird's patience, and after a time of lying still they saw their tormentors fly away and were able again to go about their affairs in the tree tops.

Birds other than the jays and woodpeckers probably have just cause for complaint because of the activities of the squirrels about their haunts, but few of them are able to show their displeasure so conspicuously. I do not believe a gray squirrel has any conscientious scruples that would cause him to hesitate a moment in taking eggs at any time if he wanted them. But in spite of the fact that they are mischief-makers and thieves in relation to the birds, I feel free to plead reasonable protection for these amiable and fascinating rodents. Their extermination, or even great reduction in numbers, would be a distinct loss to our faunal life and would greatly diminish our enjoyment of the coniferous and oak forests. No animal seems so much a part of these woods and few creatures give us such vivacious pleasure while under observation. Their lively cheery chatter, their impulsive explosive barks, are as comforting and infusive of joy as the songs of many of the mountain birds.

When early spring days arrive, the squirrels are en-

DENIZENS OF THE MOUNTAINS

gaged in the activities connected with the birth and rearing of the young. Nesting sites are sought and the snug chamber for the baby squirrels is prepared. Sometimes the nursery is made in a hollow tree, but as often a special globular nest is built. This last is generally placed some forty feet above ground in the crotch of an oak or on the spreading limb of a pine, and consists of small branchlets compactly woven together into the form of a hollow ball about two feet in diameter. First a platform is made (sometimes an old jay's nest is used as a base) and on this the superstructure is built. A circular opening on one side leads to a capacious inner chamber about eight inches in diameter which is well lined with soft shredded bark and feathers.

The young, which are born as early as April and as late as August, are two to four in number and remain blind for several days after birth. They are almost hairless, but nature soon covers their nakedness, and the little tails early give prophecies of the fluffy, bannered appendages of the adult squirrels. Their early days are spent largely in sleep and nursing, but as soon as the young squirrels are able to run about play becomes the dominant occupation.

In the latter part of June of the year 1927 my attention was called to two juvenile squirrels living about a large mountain oak. They were fuzzy little fellows with faces comically innocent. At almost any

THE ANTHONY GRAY SQUIRREL

time during the middle of the day they could be seen frolicking aimlessly about on the tree, chasing each other, or ridiculously biting and pulling each other's tails. Often one came up to the other in an almost belligerent attitude, but always instead of quarreling they touched noses as if to make peace. Then, almost like a little child, one would pat his playmate caressingly on the head with its fore paw.

The principal arboreal playground of these squirrels was a huge almost horizontal oak limb about twenty feet long. This they seldom left during the first days I watched them. Early in the morning the young ones were always in a particularly frolicsome mood. Sometimes one, as if at a loss to know how to better engage his time, would go out to the end of the limb and there amuse himself by vigorously chewing his tail. Or again, the two would chase each another around and around the tree after the manner of chipmunks playing tag. Finally meeting unexpectedly face to face and touching noses they would place their heads side by side and with little gray chins pressed to the limb peer curiously down at me as though inviting me to say, "What good little children you are!" How those brilliant laughing black eyes did glisten in the sunshine! It seemed sometimes they were made of purest jet. Between times of frolic, the babies exercised their jaws on the hard bark, or reached out and bit off leaves from the tiny branches that they might watch them float like dancing fairies

DENIZENS OF THE MOUNTAINS

to the ground. All this exuberant frolic as well as the more gentle pastime was carried on as quietly as could be. It was about the most beautiful play I had ever seen.

During the day the little squirrels made slight use of the chamber that was their sleeping place at night. If siestas were indulged in they were generally taken out on the trunk of the tree. Frolicsome and vivacious as they were these youngster squirrels could not always frisk and play, and much time was spent in resting, especially during the middle of the day. Generally they lay stretched out with the chin and abdomen pressed flat against the top of a horizontal limb and with the tail left dangling that they might enjoy feeling the breezes whisking through the fuzzy banners.

For the most part the mother squirrel left the little ones to take care of themselves. This they did very well. As though in obedience to a maternal command, they never descended to the ground where they were exposed to the attack of terrestrial predators. The appearance of a crested jay and proclamation of his arrival by a series of discordant ironical cries sent the youthful squirrels scrambling like animated jumping jacks into the home hole.

The young of the gray squirrel remain with the parent until about two-thirds grown when they begin to shift for themselves. I noted this pair was still about the home tree late in September.

THE ANTHONY GRAY SQUIRREL

As early as the first of June the adult gray squirrel is into the cones, impatient until the nuts shall be fatter and riper. By August he is in the midst of the harvest and before the end of the season the evidence of his labors is seen under almost every cone-bearing tree. The plump nuts of the sugar pine seem to be his first choice, then come those of the yellow pine. These are easily procured and may generally be had in abundance. The largest nuts are found in the Coulter cones but he finds these nut chests hard to enter. The scales are hard, thick, and gluey with pitch, and the cones are enormously heavy to handle. He does not altogether shun them but changes his method of extracting the nuts. Instead of beginning at the bottom of the cones and biting and pulling off scale after scale, he gnaws a hole in one side of the cone and then gets the nuts from the inside. Doctor Merriam noted that the Richardson red squirrels of Idaho used the same clever method for obtaining the nuts of the white-barked pine.

With the ripening of the acorn crop, the gray squirrel puts on the old fashioned jacket of thrift and begins to store food in earnest against the days of winter scarceness. Tree hollows are filled and caches made in the crevices of bark. Even so, he is not so provident as his eastern cousin, the red squirrel, and his nut supply would seem altogether inadequate in the eyes of that sagacious rodent.

We can not but marvel at the ability of these squir-

WHEN SNOWS ARE DEEP EVEN GRAY SQUIRRELS SOMETIMES ATTACK THE BARK OF THE PINES

THE ANTHONY GRAY SQUIRREL

rels to digest such tannin-filled food as the acorn. The California Indians, whose stomachs are notably strong, find it necessary to leach the acorns for several hours before using them. Cattle and sheep, I am told, suffer greatly from tannic acid poisoning when forced by lack of other food to subsist largely on acorns and oak leaves.

CHAPTER VII

THE MOUNTAIN WEASEL

I USED to wonder why John Muir always went to the mountains alone. But now I know that it was about the only way he could see things. You must free yourself by several days from the chatter of company before you can adjust yourself sufficiently to get into the quiet, mystical, unhurried moods of nature. Frequent conversations are as fatal to observations as is much movement. Short journeys taken alone are usually on this account most productive of valuable nature knowledge.

The best time to make observations I find to be in the early morning, at late afternoon, and evening, though no part of the day is to be slighted. The mind and the eye are most alert at the beginning of the day. Birds are then unusually active. If you go abroad at dawn many of the nocturnal mammals will yet be out though they are then preparing to go to their lairs. After the night's inactivity the squirrels and chipmunks feel themselves endowed with new forces, and keen appetites urge them on to such expenditures of energy as are not apparent in the middle of the day.

The early morning hours are good for saunterings, but as the day grows warmer one feels the inclination to

THE MOUNTAIN WEASEL

sit still and watch. I like to take some book worthy of careful reading, go out under the trees and then take my own good time at reading, musing, watching wild ways, and recording the results of my observations. On such occasions I find my note book growing unusually fat with valuable notations, and generally on returning to camp I feel that I have seen more to give me pleasure than when I have strolled over long stretches of trail. There is a charming restfulness to such vacation hours not to be obtained on the hurried journey.

It was while experiencing one of these periods of enjoyable laziness that I got a "sit in" on the family life of a certain mountain denizen that I could not have had under any other conditions. I had gone up Woods Creek about three miles and came to a lovely little Sierran meadow set between great cliffs and hedged in by thickets of quaking aspen and willow. Its center was green with dainty grasses and its margins flecked with myriads of splendid tall-stemmed flowers. The birds, the blossoms, and I had it all to ourselves. That's what I liked best of all. I stretched myself out on the turf amidst the greenery and floral color, and I felt myself really a part of wild nature. How much better we can enjoy flowers when we get our heads at the level of theirs! Too often we look down at them and not out with them, and thus we never share fully their life of happiness.

In a willow tree close at hand was a chattering family

of fuzzy young pewees now disporting their first feelings of independence. There were three of them and often they came near me and sat all in a row on a little limb. They edged very closely together as if to enjoy yet awhile that good comradeship they had had in the nest and which they perhaps realized must now soon come to an end.

Hard by the willow thicket was a great pile of fragmental granite, the end of a large talus slope extending from the throat of a gorge that reached back as an enormous rift in the precipitate cliffs of the canyon wall. As I was watching the happy bird trio, my eyes caught sight of a movement among the rocks which quickly diverted my attention to the activities of another little family upon whose doings I could hardly look with such feelings of sympathy or pleasure. There racing in and out among the rock tunnels was a mother weasel with two of her little ones. That they were bent on wickedness there could be but little doubt. She had these youngsters out on one of their first hunting expeditions and was schooling them in the hard-hearted business of carnage. In their greedy blood-thirstiness they were ransacking every cranny and crevice of the rock pile for such small rodents as they might find hiding in refuge there. On they went, going under one rock ledge after another, only reappearing from time to time to crane the neck and look about before going in undulating running gait to continue their slaughtering.

THE MOUNTAIN WEASEL

Though the parent and youngsters hunted together it seemed quite evident that each was out for its own quarry. Once engaged there was no thwarting of their purpose, no changing of mind in the pursuance of their helpless prey. Many times they crossed and recrossed their tracks as if to assure themselves that not one of their timid neighbors should have missed being routed from his retreat. Eventually I lost sight of the prowlers, but after a little while I saw the mother appear at the opening of a crevice with some small animal in her mouth. It was as limp and lifeless as a bunch of rags. As best I could make out it was a cony. The weasel dragged her booty across a number of rocks then quickly made off with it to the adjoining aspen thicket, where doubtless the whole gang of cut-throats had a meal of it.

With such neighbors about I could hardly see how the old pewees had successfully raised their brood, for the weasels are well known enemies of nesting birds, their eggs and young. The fact that the wood pewees usually build fairly high in the forest trees may have accounted for the early success of the attempt at raising young. But when the awkward, unsuspicious birdlings were down in the brush, life could no longer be so safe, and as I left the meadow that sunny afternoon I could but fear for the future careers of the young birds. Adult birds seem to recognize in the weasel an hereditary foe, and I have had occasion in another chapter to speak

DENIZENS OF THE MOUNTAINS

of how cordially they hate and how seriously they dread this bloodthirsty creature of opprobrious habits.

Yesterday a weasel went through the camp and all day the mountain chipmunks and golden-mantled squirrels manifested much uneasiness. They did not come into camp nor go out to their feeding places with their usual frequency. They emerged from their burrows but only to give their alarm notes. Late in the day I found a chipmunk in the trail with a hole in the side of his neck, the victim of mustellid cruelty. These little creatures were well aware of the danger that lurked near, and they doubtless realized how defenceless they were against the attacks of this bloodthirsty neighbor.

Mountain weasels that I have had in captivity have always been ready to give manifestation to their surly untamable tempers. They are quick to spring and will when provoked give short, sharp barks and hiss and snarl in defiance. Several times I have heard them give a harsh chattering note. For the most part though they are quiet little beasts, too sullen perhaps to speak except in rage and cruel contempt.

When in April or May the female elects to take upon herself parental duties the nursery is made in a hollow log, a rock crevice, or in the burrow of some animal which perhaps the thief has evicted, and there from four to eight young are brought forth. The little ones are said to be blind for nine days. Not until they are

THE MOUNTAIN WEASEL

six or eight weeks old does the mother take the troop of young pirates forth to learn their lessons of carnage.

The distribution of the mountain weasel is an extensive one. It is known to occur in California from the San Jacinto Mountains north through the Sierras as far as Mount Shasta. It is found also in the San Francisco Mountains of Arizona (where the type specimen was taken) and north through the Rockies to British Columbia.

CHAPTER VIII

THE NUTHATCHES

WOODS and mountains without birds would be without romance. No tree is so beautiful or so full of interest as one with birds in it. The beauty of the vegetable world is in itself not enough for us. We must see with it the high beating life of birds and beasts. Even an old dead stump has its attractions if it is inhabited by a family of birdlings. I now recall with what meaning the remains of a certain wind-broken pine came to be invested when I found it to be the abode of a company of those happy wood-dwelling birds, the nuthatches. It stood about twenty feet high; devoid of bark, weather worn and full of holes, for the white-headed woodpeckers had used it as a tattooing place season after season. In fact, the tree in itself had little of beauty about it. Every evening at about the hour of sunset when the meadow grasses and trees were bathed in the mellow yellowing lights of declining day, a little company of pygmy nuthatches came to flit playfully about its decaying top, while chattering and busily talking over the adventures of the day. After a short half-hour of beautiful play, they all went to bed inside the old tree and would be seen no more until early dawn when they again came out to have a little

THE NUTHATCHES

season of familiar frolic and visiting. And then, seemingly as if at a signal, the whole chattering company would whisk away and go trooping from tree to tree through the woods, perhaps not to return again until evening. Since I had my bed made on the pine needles immediately beneath the place, I could accurately observe all the movements of the birds. I have seen many little flocks of nuthatches about many old tree stumps but none has endeared itself to me as has this one.

Nuthatches are pleasing little birds wherever we see them, bright and merry forest midgets, always doing a service of distinct value to the trees. Their vegetable diet is such that they cheat no hungry human mouths and their habit of eating spiders, beetles, moths, and caterpillars, and ants and wasps marks them out as birds worthy of commendation and protection.

When feeding, the pygmy and red-breasted nuthatches are given to traveling in noisy little family parties with chicadees and, strangely, often a kinglet or two is found tagging along. The slender-billed nuthatches are likely to be hunting alone. All of them may be seen flitting, scrambling, and nimbly creeping about on the limbs of the forest trees hunting diligently for insects. They go over the bark crevices of one tree after another, deftly prying off the scales or pecking deeper to lay bare larvae hiding beneath the surface. As they chisel into the bark or decaying wood the nuthatches swing the whole body as on a pivot to give effect to each stroke, and though we would hardly think it possible, they are able to pierce the shells of nuts with fairly heavy covers.

There is always a tendency among the birds when hunting over the pine tree trunks to ascend or descend the tree vertically for a short distance and then to sidle a little before continuing the upward or downward journey. This course is probably determined by the position of the crevices. When hunting on other trees or on the smaller twigs of the pines they spiral promiscuously about. They always hold the body closely appressed against the bark and this gives them much of the appearance of some of the smaller woodpeckers. It is not surprising that the uninitiated in bird lore think of these bark frequenting birds as being very close to the woodpecker family.

The nuthatches are birds of industrious habits and "as if to become extra healthy and wealthy and wise" are at work early in the morning, often before sun up.

Though the nuthatches are very little birds, they, like the chickadees, raise big families, and doubtless this accounts for the great numbers we see. The mating season is early May, and the birds are then usually restless and may be seen chasing each other from tree to tree. Finally pairing off, they go bustling about hunting for nesting sites.

Ready-made cavities are very easy to find, for the industrious woodpeckers hollow out so many holes

THE NUTHATCHES

that it would seem the nuthatches' greatest difficulty would be to make a choice of location. But most of the woodpecker holes are not suited to the nuthatches' needs, being entirely too large. Only the holes of the sapsuckers, it seems, might prove attractive. The nuthatches are quite able to excavate and generally do make a cavity for themselves, providing they can have partly decayed wood in which to work. If, as is sometimes the case, the little worker meets a hard section of wood she pluckily tackles it, and with persistent and untiring blows pierces as much as is necessary for the completion of her home. The cavity prepared for the nest is sometimes made in old stumps and not more than four or five feet from the ground, though as often a high site is chosen. The hole does not generally extend far into the tree nor is it often deeper than five or six inches. The cup-shaped bottom, two or three inches wide, is filled with little pieces of rotten wood and upon this is placed the nest proper, consisting of plant wools, soft bark fibers, cobwebs, feathers, and a goodly quantity of rodent fur. It is interesting to speculate as to the manner in which these little birds get the rodent fur that they use for the nest lining. They are not adept enough on the wing to get fur from the backs of the lively squirrels, wood rats, and chipmunks, as the imaginative would have us believe, so we must conclude that they find it bit by bit where it was dropped by shedding animals, or else that they

DENIZENS OF THE MOUNTAINS

come upon it about rodent's sleeping quarters, or where it has caught in the crevices of the rough bark. If we look about old woodpeckers holes, knotholes and the like, often we will find ample evidence of their having been occupied in the past by rodents of various species. All around the edges of the openings are hairs that have become fast in the clefts of the bark, and the nuthatches would have no difficulty in finding them, for their travels over the bark take them into just such places. It has been suggested to me that owl pellets may prove to be the source of some of the fur, but that they supply any considerable amount is doubtful, for the pellets are not anywhere plentiful enough.

After the young have been reared, the nuthatches again join in noisy troops and journey through the woods in random manner. The frequently uttered notes are difficult to describe, but among none of the species do they partake of the nature of a melodious song. The various whistles and peeps are uttered so earnestly that they seem to be conversational.

Three species of nuthatches with their various subspecies inhabit the mountains of the western United States, the slender-billed nuthatch (*Sitta carolinensis aculeata*), the red-breasted nuthatch (*Sitta canadensis*) and the pygmy nuthatch (*Sitta pygmaea*). They are all compactly built little birds with bluish gray backs, slender compressed bills, and short stubby tails which they never use as props as do the woodpeckers.

CHAPTER IX

THE WHITE-FOOTED MOUSE

WET seasons bringing abundant harvests of native seeds are always good years for the wild mice, and a series of such seasons may bring an unprecedented increase in individuals. The little rodents become very noticeable, especially about human habitations, and no sooner does darkness come on than we see the dainty creatures running about on the sills of our shanties or scampering across the floor in search of crumbs. The white-footed or deer mice are so unusually beautiful that, were it not for the fact that they somewhat reveal in their form and habits kinship to the common house mouse, we would not wish to molest them no matter how abundant they became.

During a recent summer when I spent a number of weeks as the guest of a friend at a mountain resort, I thoughtlessly placed my ties and handkerchiefs in a bureau drawer that was none too mouse proof. Not two nights had passed before a white-footed mouse found the fluffy silks and linens, realized their possibilities, and chewed them into tiniest shreds. She hauled in pieces of cord, feathers from my pillows, and cotton from my mattress, and soon had the materials made into a snug and cosy nest.

DENIZENS OF THE MOUNTAINS

Did I give way to wrath at the sight that came to my eyes when I opened the drawer and saw the ruinous tax that had been levied by this mouse upon my clothes? Did I call for revenge? Was I thirsty for the blood of that little creature that now fled terror-stricken from the nest and tremblingly peered at me from out those great black knowing eyes as she sat frightened on a sill far above my reach? A moment's work then with a broom, or with a trap that night, and the limp body might have lain dead in my hand, never to molest me again. But no, there was a better way, a saner choice, a different attitude to assume toward her if I would be wise. I would let her live, follow her activities, and know of her ways with her children. Only living mice can contribute store to our knowledge of rodent behavior. I reasoned that, had I been a mouse, I would have done just what she did. The law of my nature would have led me to select for a nest just such soft materials as she had. There really could be no blame, except upon my own head. I should have known better than to leave my things exposed to small rodents. So now I gave over the drawer to the little mother. I did more, I put food out for her and took kindly interest in her welfare.

A few days later four small, pink, blind and helpless youngsters were discovered in the nest, which had so perseveringly been built. During the day the mother remained dutifully with them, but always at dusk she

THE WHITE-FOOTED MOUSE

covered them over carefully with the materials of the nest and went out to feed that she might supply milk for her family. If I disturbed her while she was nursing the young (they remained attached to the mammae most of the time), she fled from the drawer with all the little ones tenaciously hanging to her teats. It was highly amusing to see the diminutive creatures being rolled and bounced about while hanging for dear life to their fleeing parent. The cracks and bumps they received as she flew along must have been many, but they made not the least audible objection. The little mother had all her wits about her even in flight, and once when one of the babies lost his hold she went back to him, took him by the skin of the neck, and went on her way out of the drawer, up the tent wall, along the sill, and out-of-doors.

Once, after such a hurried flight to safety, she decided to change the location of her nest and set up housekeeping in the top of the burro's barley bag—in the eyes of a mouse, a wise and promising site and conveniently near the pantry, I should say. But in reality a home placed in such a location was more insecure than one built in a bureau drawer could possibly be, for twice a day I had to get grain from the sack for my animals.

In the meantime, some little children came running to me and excitedly exhibited in a box, so they said, "orphan mice" which they had found in a hollow log.

DENIZENS OF THE MOUNTAINS

Inasmuch as I had let it be known that I was interested in wild mice and had regard for them there was nothing for me to do but take them. I tried feeding the babies with a medicine dropper, but though fairly well grown and covered with dark hair they would not take the milk. I did not want to kill them and so in desperation I went to the barley bag where my other mice were, clapped quickly a tin box over them, and transferred the mouse-mother and her progeny to a square five-gallon oil tin, from which I had removed the top. Then I put the "orphans" in too, covered the can with a fine wire netting, and watched to see what would happen. The strange new mice were no sooner inside the can than they were noticed by the mother of the others and adopted by her. They were all dreadfully hungry and she allowed them to suckle in preference to her own young. This was late in the afternoon.

Next morning I went to observe my mice and to my horror all that remained of them were the half-eaten remains of two little ones. It was my first thought that a skunk had taken them, mother, babies and all, but a moments's reflection told me that this was an incorrect conclusion, for the screen I had put over the top the night before was still in place. There was a small hole in the netting, but how could the mother mouse escape through this when she had to go up to it over the smooth sides of the can?

I went to the bureau drawer where the original nest

THE WHITE-FOOTED MOUSE

had been, just to assure myself that the mice could not possibly be there, when behold I found not only the parent but also the little "orphans" by her side. Evidently she had not liked the tin-can house and, after having given way to cannibalism, left with the better-liked adopted children for a place more to her choice. I am still left to guess how she got them up and through the small opening in the screen. Maternal instinct often solves problems that ordinary human reason would declare impossible. Knowing how well mice can jump, I am inclined to think that she leaped up to the screen and then wandered about on the underside of it until she found an opening.

It will perhaps disgust some of my friends who are economic zoologists to learn that I allowed this mouse family eventually to escape from the nest and to wander out into the forest and increase its kind. From the standpoint of the agriculturist and the forester, I am fully aware that on the whole such rodents are a great nuisance and should in most instances be destroyed, but familiarity with this little family and acquaintance with its trials made my heart soften in its favor.

Mice of the genus *Peromyscus*, to which the creature of our sketch belongs, are plentifully found almost everywhere in the mountains, and often the collector will be able to find three or four species living in a small district. Osgood, who made a special study of mice of this genus, tells us that "throughout practically all

Photo by Donald R. Dickey

WILD MICE ARE COMMON FOREST NEIGHBORS

THE WHITE-FOOTED MOUSE

of the western United States they exist in countless numbers, perhaps exceeding those of the other combined mammalian inhabitants of the region." It is their custom in the natural state to live in holes in the ground, in rock crevices and in decaying stumps of trees. They are wholly nocturnal, but they have no objection to the light of a lamp or campfire. They are most graceful and fairylike, and it always has given me pleasure to watch them about my camp. They have helped me while away many a pleasant hour when I was all alone.

In one of my permanent shanties I have a hurriedly made cupboard. The two doors were for some time fastened by a simple string catch and they had a tendency always to fly open when the cord was untied. This cupboard contained my small supply of groceries, and naturally was early discovered by the keen-scented, ever-hungry mice. Every evening when I was there I could see the soft-furred, white-footed mice peaking out from behind its corners as soon as the light was lit. So long as I was about they were very cautious, and doubtless if I could have read their thoughts I would have found that they were always glad when the late hours came and "that big animal who makes light at night time and who sits so long staring at white papers" went to bed and left the house at their disposal.

One morning, soon after the cupboard was made, I found its doors wide open and the unmistakable evi-

DENIZENS OF THE MOUNTAINS

dences of mouse nibbles on the cracker box inside. It was my first thought that I had in forgetfulness left the doors unfastened the night before, when I put away my food after the evening meal. Examination of the string catch, however, showed that it had been broken in two. This, I said, is because it was too weak and it gave way under the strain of holding the doors shut. So on the following night a new cord was used. In spite of this I found next morning that my doors were wide open again, and the string in two as on the previous day. When the same thing occurred three nights in succession I began to suspect that it was not accidental, and my thoughts turned to the innocent-looking little deer mice that peeked at me so often from behind the cupboard. But against them I had only circumstantial evidence. I now decided that even though my groceries were in jeopardy I would continue using a string as a door fastener and attempt to get unmistakable proof. Successful thieves get bold, and on the very next night the culprit faced the lamp light, came out on the top of the cupboard, and before my eyes "snipped" the string. The doors no longer held by the catch opened wide and in went the little robber! Had the mischief maker learned the certain relation that existed between the gnawed string, the opening doors, and the good things to eat inside? Evidently so, but I will leave it to some one better versed in mouse psychology than I am to explain the steps by which the

THE WHITE-FOOTED MOUSE

small, rodent brain reached this attainment in learning. A steel chain now does the duty of the cord.

The white-footed mice are rather indiscriminate feeders and vary their diet according to the season and the available food supplies. Their cheeks are pouched and the mice are broadly resourceful in finding little seeds to fill them. The seeds of composites and of grasses are much relished and at times small supplies of these and other foods are stored as a stand-by in hard times. Insects form no inconsiderable portion of the diet, and meat is always readily consumed when it is their good fortune to find it. These little sharp-toothed marauders prove themselves a constant annoyance and hindrance to the professional collectors of mammals, for many of the small animals caught in traps during the night are eaten by them before the carcasses can be removed for skinning.

Though these mice live in the mountains up to very high altitudes they are nowhere dormant during the winter time. During inclement weather they may be forced to remain under shelter for a time, but it is not long before their tiny tracks may be discovered on the snow again. They are good tree scalers and nose around in the cracks of the bark for food when it is not available on the ground. Sometimes their journeys made in search of food lead them into most perilous places, for when they are in exposed situations the ever hungry owls, weasels and skunks easily make way with them.

DENIZENS OF THE MOUNTAINS

This morning as I walked down the trail I met an elderly lady friend with a botanical collecting can in her hand. As we chatted concerning the plant specimens we had taken, the drift of the conversation led her to tell me of a white-haired gentleman she had seen but an hour before and of the experience they had in helping a perplexed little rodent in peril.

"When I met the man," said my friend, "he was looking down into a deep, narrow hole made to hold a telephone pole. 'See what we have here,' he said. I looked down some five feet to the bottom and there spied a little white-footed mouse looking up bewilderingly and pleadingly at us. How long he had been there we could only guess. The kind-faced old gentleman, with a lively feeling of compassion, put a stick into the hole, and as though the tiny creature realized the purpose of our endeavors it immediately crawled up the pole to ground level and then hesitatingly jumped into the the grass. We could but marvel that he did not seem more afraid of us. Perhaps he knew he had a friend."

A simple incident is this but it made a deep impression on my mind as I listened to its narration, and I will long remember that gentleman whose regard for the feelings of a tiny mouse led him to put down a stick into a hole that the "cowring', tim'rous beastie" might be rescued from starvation, and again have the freedom of the fields. Such a man is one of God's true noblemen.

CHAPTER X

THE CLARKE NUTCRACKER

THE nutcracker is among the handsomest of our crows. His livery shows no color combinations of questionable taste, but pleasing and pronounced contrasts of modest hues. The attire differs from that of all other American crows in being for the most part a lead color disrupted by greenish black and white on the wings and tail. The strong beak as well as the feet are black.

During summer the high ridges forested with lodgepole pines are his favorite habitat. As far as conspicuous birds are concerned, Clarke crows often at this time hold undisputed possession of the high forests. At times when I have camped for protracted periods in the boreal woods I have become quite weary of their harsh, raucous cries and would have welcomed with eagerness the song of even one fox sparrow just to have the monotony broken.

Wherever we see these mountain crows we find them to be exceedingly entertaining birds. Their inquisitiveness knows few bounds. Some are always loitering around and as soon as they see the traveler entering their domains they, like the jays, set up a vociferous chatter and give a series of strange nasal cawing notes

Photo by L. M. Huey
CLARKE NUTCRACKER

THE CLARKE NUTCRACKER

to let their companions know what is going on in their forest home. While keeping the intruder in full view, they are always judicious enough to remain at a safe distance. It is generally their habit to use the tall dead trees as observing posts, and as they sit upon the bleached and weathered tops their plain colors blend so perfectly with their surroundings that they are likely never to be seen. As the trees from which they are watching are approached the birds launch forth on strong wing beats and fly quickly to others just ahead, setting up in the meantime loud, unmusical, complaining cries of protest, a *kr-a-ck, kra-a-a-ck, kra-a-ck,* which sounds much like the cry of a hen when she is being carried to the chopping block!

After the birds have become accustomed to the camper's presence their inquisitiveness leads them closer and closer to the objects of their curiosity and eventually they become very bold in entering the camp premises. They have been given the rather opprobrious name of camp robber, and I am afraid not altogether without reason. Observation of their habits soon brings to light that well known racial trait of the crows, the desire to steal. Bread crumbs, meat scraps and cheese rinds are always deemed to be morsels too dainty to pass up and they will often maneuver for long periods to get into camps undetected. Though they at times appear to be somewhat of a nuisance yet hardly can they be called birds of annoying habits.

DENIZENS OF THE MOUNTAINS

They always show enough gaiety and good fun to make us feel that we can forgive their bad manners.

They continually give us the impression that they think their mountain home is a pretty good place in which to live and that we were sent in their midst with provisions to give them a little variety of food and pleasure.

Though this bird was known to ornithologists since the early part of the nineteenth century, almost seventy years elapsed before any nests were obtained and even to this day specimens are extraordinarily rare in the best collections. This is to be accounted for by the fact that the birds breed unusually early and that exploring parties, because of the danger of inclement weather, seldom enter the mountains until after the birds have reared their young.

It was a long journey I made to get my first sight of a nutcracker in its nest. Several years ago when on the return home from a trip into Death Valley, I chose a route through the Amargosa Desert which led me past a mountain range of unusual beauty and uniqueness, the Spring Mountains of southern Nevada. The very remoteness and wildness of these forested highlands, far out on the untenanted deserts, recommended them to me as a place worthy of study, and I felt certain that if I could but spend even a few days exploring there I should be one of the happiest of mortals. The desire of the years was satisfied when during March, 1928, I

THE CLARKE NUTCRACKER

journeyed across the broad Mohave Desert enjoying its wide expanses of primitive wildness and I came again to the Vegas Valley in southern Nevada along whose western borders I could see my noble mountains standing waiting in solitary repose and grandeur. The country all about is so strewn with peculiar mountains of rugged form and moderate elevation that one is apt to misjudge the height and majesty of this fine wild range of which Charleston Peak is the outstanding eminence, rising almost 12000 feet above sea level. For twenty-two miles the road led me expectantly up the detrital cone, which gradually rises up to the timber. Tree yuccas, cactuses, and other xerophytic plants were my plant companions almost to the 6000-foot level where their places were taken by small junipers and piñons. Then came the forest of yellow pines and firs. The canyon up which the road passes ends in a large amphitheater surrounded on three sides by great conglomerate cliffs almost three thouasnd feet high. So steep are most of these cliffs that only a few trees are able to maintain growth upon them and the forests for the most part occupy the valley floor. The precipices are at once so grand and the rock forms so sublime that it takes days of living in their presence for one to fully realize their true nobility.

As I entered the mountains proper I saw at once that they offered a likely place for the abode of Clarke crows. The great expanse of piñon forests is naturally attractive to these birds during the nutting season and the high forested ridges offer them just such summer quarters as they like. A few days before my arrival a storm had blown eastward from the distant Pacific and now a deep blanket of snow mantled the entire range. But this did not deter me from living in the open and with the aid of my companions a camp was made beneath the pines. Though the nights were exceedingly crisp so that water froze solid in our vessels, yet the days were delightful. The first morning after my arrival I saw several nutcrackers about but they were all as silent as could be. This I took to be good evidence of their nesting activities and as I wandered abroad in the forest I kept a sharp lookout for nests. After several hour's search I was much gratified to see a crow flying out of a fir tree almost in front of me. I was coming down hill at the time and could look right down into the tree. After scrutinizing the higher limbs for only a few seconds my eye fell onto the rather bulky nest. So extraordinary was my good fortune that I could scarcely help thinking myself enchanted. If I were even a half expert at tree climbing I could have had a good look into the deep cupped nest, but as it was I had to content myself with a somewhat superficial view of it. To confirm myself that the nest was no other than that of a nutcracker I waited until the bird returned and occupied it.

THE CLARKE NUTCRACKER

With snow on the ground and the nights so cold (the bare remembrance of them now almost makes my teeth chatter), it seemed a queer if not indiscreet time for a bird to attempt the rearing of young. What a frigid nursery! Evidently it is the belief in nutcrackerdom that nesting duties should be disposed of early in the year and that children should be inured to cold from the very beginning.

When next morning we were driven from the mountains by a storm of sleet and snow I could not but keep my thoughts on that old mother bird sitting on her nest during such inhospitable weather. Doubtless she was wholly unconcerned and took it as a part of a fore-ordained program.

When September comes these birds of gregarious instinct assemble in very considerable flocks of a hundred or more and go down into the yellow pine and piñon country to feed on the ripening pine nuts. This is evidently the time when the nuts are stored against the day when they will be needed to feed the young. In the extensive upland mesa region of northern Arizona the birds are in autumn very noticeable about Williams and Flagstaff. Whole troops of them may be seen flying about and heavily alighting like black crows in the junipers and piñons. Life now seems never to have a dull moment for them. They are constantly moving about crying to one another in friendly concourse while busying themselves among the cones.

DENIZENS OF THE MOUNTAINS

On one occasion I was resting under a juniper tree when a large company of nutcrackers settled to feed in a group of piñons just to my right. While most of the birds were too much occupied in the harvesting of nuts to be aware of my presence, three spied me and came especially near. But after looking on me with sort of an idle curiosity, all but one flew hastily away to join their noisy companions now busy in other trees.

This bird of marked inquisitiveness would not be one of the crowd. He must reconnoiter my appearance more closely even if he were deserted by his companions. Having satisfied himself that I was harmless, he began leisurely pottering about the camp and soon discovered that it was a good place to pick up crumbs. Almost immediately he began to gabble something as if he were talking to himself. Then he set up a vigorous harsh-voiced cawing which I interpreted as a call to lunch, for within a few minutes a dozen or more of his comrades were there with him. Nor did they leave until they had taken care of every morsel of the ungrudged feast.

CHAPTER XI

THE BELDING GROUND SQUIRREL

ALONG many of the same High Sierran meadow borders occupied by the golden-mantled ground squirrels there are often seen small rodents showing a markedly cousinly connection with the prairie-dogs of the Great Basin region. These are known to naturalists as Belding ground squirrels, but to local residents of the mountains as picket pins, because of their habit of sitting bolt upright about their burrows.

These plain-colored rodents are much less known than other ground squirrels because of their high and restricted mountain habitat. They occupy territory almost wholly unreached by the automobiling tourists and the foot travelers who penetrate to the high interior meadows are as yet comparatively few in numbers. They have had good luck above most of their cousins in choosing for their home a situation where their burrows and food habits would be the cause of no concern to the agriculturists.

The traveler first becomes aware of the presence of these animals when he hears before him the sharp, forceful, bird-like utterances of alarm sounded by some alert and watchful squirrel before it dashed into its hole for safety. Such a warning is a signal for all other

GROUSE MEADOWS, HIGH SIERRA, HOME OF THE BELDING GROUND SQUIRREL.

THE BELDING GROUND SQUIRREL

DENIZENS OF THE MOUNTAINS

squirrels in the little colony to flee unhesitatingly for shelter. If the observer when walking onward sees no signs of life about him, he may be sure that he has been spied first by one of the animal sentinels and that all the other squirrels have secreted themselves in their burrows before he arrived in their midst.

It has never been possible for me when watching the little fellows duck for cover to keep out of mind the action of a gang of boys at a street fight when apprised of the approach of a policeman. The lads are generally well hidden by the time the officer arrives and he almosts thinks that his eyes told him wrong, and that there never had been a boy in the neighborhood for days in the past. Even as with the boys, the chances are good that within fifteen minutes after the intruder has passed every member of the ground-squirrel colony will be out again sitting up in picket pin fashion and looking wise-eyed while listening for further sign of danger and alarm. "Is every thing safe? No enemy here?" they seem to be asking.

One morning after I had made my camp at Long Lake near Bishop Pass I heard the Belding squirrels making merry with their whistling at sunrise. An investigation revealed that seven families of them were living in the meadow below me. The bright-eyed, sunny-countenanced young were now (July 20) about one-third grown and out everywhere feeding on the tender grasses and succulent herbs. They invariably

THE BELDING GROUND SQUIRREL

scuttled for their holes whenever I came near them, as if they half suspected themselves of being poachers in my private garden. Always did they show much less caution than the experienced adults. No more were they under cover than they popped their heads into daylight to see out of curiosity what the strange moving visitor was. Like the older ones they sat bolt upright in front of the burrow, sometimes two of them together, and with hands pressed to their chests like little fat priests, they watched and listened that nothing might escape their attention.

A parent with three young had her home under a big flat rock in the middle of the meadow. A careful investigation of the site showed a strange choice. The ground was so damp and the sod so spongy under foot that it appeared the most unhealthful of places for any animal to live. There were numerous small openings all around the edges of the rock. Indeed, it seemed so completely undermined that had the burrowers removed a few more armfuls of dirt it must surely have come down upon their heads. I took my observation position on top of the rock and from the many holes about the margin I could see first one, than another, sometimes two together, of the little black-eyed fellows, inquiringly poking out their heads like a lot of scared youngsters watching a band of gypsies pass by. The single adult, which could have been no other than the mother, thrust her head out cautiously just twice,

A GROUND SQUIRREL MEADOW

THE BELDING GROUND SQUIRREL

each time just long enough to sputter curses at me. Always she made a hasty retreat. Several times one of the young ones came out with his nose and back covered with fresh black dirt, evidencing that even at this early age he knew how to burrow.

Often in the evening I saw numbers of the young playing about and cuffing one another. Sometimes they amused themselves by crawling up on one another's back and then leisurely sliding off. Playing tag was a great favorite among the livelier sports. They all had fleas and frequently took dust baths. In doing this they would first sit up on their haunches. Then after tumbling over on one side they would dig their heads into the dust and roll over and over, like dogs at play, and make the powdery earth fly over them in the liveliest manner. Once while they were rollicking and tumbling about after this abandoned manner, a sharp-shinned hawk swooped down over them. It was amusing to see in what a ridiculously short time they found their way into their holes. The intensity of their flight was well manifested in the sharp notes uttered by them as they flopped under cover.

Their principal food at this time was the half-ripe seeds of a phacelia, the fruit of which was born on stout stems about seven inches from the ground. These the young squirrels could just reach conveniently by sitting up on their haunches. When in this position their tails were straightened out rigidly and

DENIZENS OF THE MOUNTAINS

flat on the earth behind them to serve as prop, proof perhaps that even the squirrels seem to know that the tripod is the most stable form for support.

Most of the tunnels of this colony of ground squirrels were made under rocks or in rock crevices, but some of them were in soft soil in the open meadows and where made, I suppose, by the foolish ones who realized none of the dangers of exposed homesites. All about I could find freshly made holes, some running into the earth several feet, but most of them mere pits. These last were evidently the work of the young and from their number it would appear that they were holes dug for no special purpose but were only excavations impulsively made, the childish attempts perhaps of the juniors to imitate dad and mother.

The mosquitoes were here so abundant and so persistently annoying that to enjoy even partial immunity from their bites and so be able to make even these few observations I had to tie handkerchiefs over my hands and face. But the squirrels, not degenerate from living in shelters, were unmolested by these pests. Indeed, it appeared that they were unaware that such nuisances as mosquitoes existed. Perhaps some observer may yet confer a blessing on our race by discovering the cause of their immunity.

To continue my studies under more favorable conditions I was now forced to move up over Bishop Pass and down along the Dusey Lakes to the Middle Fork

THE BELDING GROUND SQUIRREL

of Kings River. Here I was able to observe the Belding squirrels at very close range. In a family opposite my camp there was a mother and three half-grown young whose activities proved unusually amusing. The parent evidently felt that she had done her duty by her children and was now showing repeatedly her efforts to persuade them that it was time for them to forage for themselves. Each time they approached the hole she immediately pursued them with greatest vigor and drove them off a considerable distance. Once one of the youngsters arrived at the entrance of the home burrow while she was busy scratching fleas. But even while engrossed in ridding herself of so aggravating a pest, she dutifully found time to discipline her child. Suddenly she jumped at him and administered such a sharp cuff on the ear with her fore paw that he was sent tumbling away. Not feeling satisfied with this, she terminated the exemplary punishment by a zealous and businesslike vocal demand that left no doubt that she meant that he should leave the premises once and for always. When one sees behavior so purposeful and so akin to that of human beings he is strongly tempted to think that there are many activities among animals that can not be ascribed wholly to instinct.

CHAPTER XII

THE BLUE-FRONTED JAY

OF ALL the non-raptorial birds few are more unloved by other mountain-dwelling feathered folk than the blue-fronted jay. His approach to a tree is generally a signal for a rapid exit of the smaller birds of timid nature, and the more peppery tempered ones show their displeasure at his noisy and forward intrusion by tormenting him and menacing him with their beaks.

The western wood pewee, though so small, is one of the pluckiest birds to attack the jay. Several times during the nesting season I have seen a pair of these tiny flycatchers evict a jay from an oak with such bravery that I could but feel that they merited the highest praise. With much chippering and snapping of bills they went after their enemy and he, recognizing that they meant business, in each instance left the premises without much protest. One recent May morning I saw a small pewee alone attack a jay but under these conditions the outcome was not so favorable and the large "blue-front" made but sport of his pugnacious and angry assailant. As the little pewee came at him the jay began flying and flopping about in the tree in a most erratic manner. He screeched and squawked

THE BLUE-FRONTED JAY

like an insane bird and tumbled about much after the manner of a decapitated hen, always keeping just a little away from the avenging pewee. It was a ridicu-

THE INQUISITIVE BLUE-FRONTED OR STELLAR JAY

lous sight to say the least. The brave-hearted little tyrannid did not give up so soon as one might have expected, but kept after his offender until he saw that there was nothing further to be gained by keeping up the attack.

DENIZENS OF THE MOUNTAINS

At another time I saw half a dozen rough-winged swallows assault a jay. They were circling high in the air when he suddenly alighted in a tree immediately beneath them. No sooner had he comfortably seated himself on a limb than they spied him and lowered their course so as to be just above. Round and round they went, and closer and closer they came, until they were scarcely two feet above him. Whether the jay was actually frightened or became dizzy following their whirling flight, I am unprepared to say, but suddenly he gave a peculiar squawk and disgustedly got up and left. The swallows evidently were not yet through with him for they followed him menacingly through the woods until both they and the jay were lost to my sight.

During the past summer, while in the vicinity of Big Pine Lake, I witnessed a hummingbird go after a jay in a manner that showed that even these bird midgets are no cowards in the presence of the bluefront. I can not yet believe that in this instance the jay had been guilty of any act warranting so fiery an attack on the part of the hummer, but he got it just the same. As far as I could make out the hummingbird was in that irascible state of mind in which anything would make him "boil over" and when in his flights he had come onto a jay, that bird was the one that had to suffer the venting of his spleen. With that characteristic metallic buzz of anger, the hummer dashed at the

THE BLUE-FRONTED JAY

jay several times and then annoyingly shuttled back and forth in front of him as if ready to pick his eyes out. It was altogether too much for the old fellow and he too got up and left. The hummingbird followed closely behind and when last I saw the jay he was making for the tall timber across the lake.

Almost all of the smaller birds seem to recognize the jay as their hereditary enemy, and there is always a commotion and great ado among them when he comes around, especially if it is nesting time. There is probably no bird that is such a robber of their eggs. "When I see" says an ornithologist, "the damage done by the jay, I feel as if nature had slipped a cog somewhere, for as far as I can see the majority of birds have no means of protection against this arch robber."

When a jay finds a nest of eggs he generally meets a temptation too strong for his nature to resist, and calling to his fellows, who soon come to join him in the work of destruction, he proceeds to devour the eggs, or if young are in the nest, to eat or carry off the birdlings. Sad mess does he make of a nest and its contents. A yolk-besmeared bird home is all that is left for the distracted bird parents to witness when the skulking robber has left the place; and it may be that he has even torn the nest to pieces.

But this bird predator that brings misery to so many of the small feathered folk, has his troubles too. Only this morning I was convinced of this anew. Early I was visited by two inquisitive blue-fronts, which had come to spy out any crumbs I might have thrown out for the smaller birds. With their finely formed bodies and rich deep blue feather coat they certainly were handsome visitors. When a jay's head is cocked in inquisitive turn, with the great cockatoo-like crest thrown to one side, he certainly looks the part of an aristocrat. There are two long white spots at the anterior base of the crest which have somewhat the appearance of a pair of eyes, and because of this the crest assumes to us the likeness of the Periclean helmet, which, as you will remember, gave a double-foreheaded appearance and look of wisdom to the wearer.

I could not but watch these birds with interest. Their behavior was so wary and yet so bold, and as they alighted on the ground from the tree above they set up a series of varied and versatile cries that seemed as near real talk as could be. Finally a piece of cheese with the rind still clinging to it was found, and with renewed talkative cries they asserted their right to it and declaimed the discovery far and wide. So loud were their notes of glee that two other jays were soon attracted and came wheeling in to see what there might be for them to feast on. As the new comers arrived both scolding and notes of challenge filled the air, and when the discoverer of the cheese rind flew into an oak tree so that he might keep the precious food all to himself there was set up such an inharmon-

THE BLUE-FRONTED JAY

ious hubbub of squalling notes of protest that it seemed as though all the jays of the forest had met in convention.

In spite of covetous care, constant pecking broke the cheese rind into bits, and since the crumbs fell in several directions each of the greedy birds managed to snatch his share. Such a jay party as then it was! Silence was never to be had at a greater premium.

But tragedy often travels on the heels of gluttonous merriment. Even while I was watching and enjoying the sight of their merry time, there came a noise and a sound of swishing of wings that startled me. Almost before I had time to come to a realization of what was happening, a red-tailed hawk was on the ground with a screeching, agonizing jay in his talons. He, too, had been watching the merry party, but with a different interest, and had swooped down straight from the skies to make a meal off one of the birds. In another instant he was off with his prey.

That put a sudden stop to several things, as well as the ill-starred jay. The merry-making was over at least, and for almost all the remainder of the day there was never a jay voice raised around the premises.

Strange it seems that all of our beautiful forest must be the scene of such tragedies as this, one creature preying upon the other, the stronger on the weak. But such is a part of the present natural economy—a survival of the fittest, and Nature's way for the elim-

DENIZENS OF THE MOUNTAINS

ination of the unfit. Hard is the onward struggle of the living world.

The jay is one of the first birds to see you and pay a visit when a new camp is made. His inquisitiveness is boundless. He considers it to be his duty that little shall go on in his forest home that he does not see. "Where did you come from? What is your name? How long are you going to stay?" A dozen other inquisitive questions you can easily imagine him to be asking you as he and his companions sit inquiringly above in the trees and watch your every motion. If you move toward him the chances are that he will not fly away into another tree but that he will hop upward with great striding steps from limb to limb until he reaches a place he considers safe.

If jays are given the least encouragement, their boldness will prompt them to become familiar very soon. They enjoy a mixed diet and it is counted a stroke of good fortune when a camper moves into the neighborhood who is thoughtful enough to throw his edible table crumbs to the wild things. Meat scraps are doubly appetizing to them and every bone thrown out is sure to be picked clean.

The natural food of the jay is acorns, pine nuts and other seeds, and in addition such animal food as is available and pleasing to his fancy. In autumn we often see acorns dropped before ripe from the high branches of the golden oaks. At first I was inclined

THE BLUE-FRONTED JAY

to think it was the squirrels that pried them loose. But now I know that the work must in a large part be credited to the jays. The birds get hold of the stem just behind the fruit and then most vigorously shake the head back and forth until their prize is torn loose. If the acorn drops to the ground it is not always retrieved. It is too simple a task to find another.

During the summer the "blue-fronts" are common enough in the mixed oak and coniferous forests, but as we journey higher into the belt of firs and lodgepole pines they become rarer and rarer, until finally at about ten thousand feet and above they are seldom seen. In autumn and winter these attractive birds are sometimes encountered in numbers in the brush of the Upper Sonoran Life Zone. They are, however, typically Transition Zone species and as the lower edge of that belt is reached they begin to be displaced by the brighter-colored California jays.

The mating season is May and June and soon thereafter jay's nests of the season may be seen in the pines, cedars, and firs. The rather bulky structure is placed high—twenty-five to fifty feet— and is constructed of small twigs and grass. The lining is usually fine, thread-like roots, and the whole is held together by a daubing of mud. The eggs, which number from three to five to the set, are pale bluish green, freckled with brown and lavender.

DENIZENS OF THE MOUNTAINS

By July 15 most of the jays are through nesting and family parties of old and young may be seen in the trees. During the nidification the adults are very quiet in comparison to their clamorous record of the

YOUNG STELLAR JAYS

late summer and winter. But the newly fledged young are noisy enough to make up for this period of comparative muteness on the part of the adults. The cries of the young for food seem unusually persistent.

THE BLUE-FRONTED JAY

The annual molt takes place in August and then for some weeks we see the jays in unusually fine clothes. As the season progresses the bright raiment begins to show wear and if his enemies have subjected him to too rough handling his coat may indeed soon appear but a "thing of rags and patches." Feathers are not living structures but dead ones, and the molt is Nature's only provision for changing from tatters to silk. The plumage of the juvenile is duller than that of the mature bird, but is changed for the adult winter livery early in autumn.

CHAPTER XIII

THE GRAY FOX

I LONG ago took to heart the advice of an early-rising prospector-naturalist, that one should get up at dawn and try to be off on the trail before all the nocturnal wild creatures have gone to their lairs for the day. It is during the night hours that most of the wild animals are out and active, and if we would see them we must at least get up an hour or two before sunrise. Among the first tracks upon which I come when on my morning wanderings are those of the gray fox. The cat-like marks are so plentiful along the sandy wash that runs alongside my mountain camp that almost instinctively I call the place Fox Alley. The country all about is very rocky and the foxes to save their feet have chosen the sandy canyon bottom for their principal highway. "Here," said I one day, "I will sit down beside the trail to see what comes by."

Now it would be very pleasing, if I were making a story, to have had a fox come by on that very first morning's watch. Nothing so fortunate occurred. I watched there beside the trail a good many early mornings and several moonlit evenings before I saw my fox. But the sight I eventually had of it well repaid my days of waiting, for under the conditions I saw an

THE GRAY FOX

animal, not running away from me in fright, but one acting wholly naturally. It was a female bound for the home lair with a wood rat in her jaws. Gracefully and proudly she trotted along, with her head well up and with the gleam of happiness in her eye. And by that sign I knew that there were young foxes at home. I watched the beautiful creature as she disappeared up the canyon in the dusk. This was one of the red-letter experiences of my days in the wild.

After that evening time I often found my place among the rocks above the trail. I generally saw other things than the fox I sought: wood rats and white-footed mice, and once a bobcat. But every few days I saw that mother fox out again, hunting food to satisfy the hungry mouths at home. To make it more attractive for her to come that way, I buried meat scraps in the sand, and soon she began to make the journey past my lookout very regularly. Often I heard, just as darkness was coming on, the single, hoarse fox bark carried out on the cool still evening air.

I can not tell you how much I wished to know where the lair was. Many days I searched, but I always came home from my hunt, baffled and unsatisfied. There was always the thought that if I did not find the den soon, the young foxes would grow up and leave the burrow and I would never see them. It was, then, a great satisfaction when after a long ramble I one day saw the two baby foxes playing about the entrance of a small hole in the side of a little gulley. I approached them so cautiously that they manifested little more than curiosity toward me when at last they caught sight of me. I kept a fine screen of brush between them and myself and I don't think that they had any realization of danger. I remained at a fair distance and then sat down to watch them. I saw one of them run playfully into the den and soon come out dragging with him a bunch of feathers. This he pawed over, took again into his mouth and shook thoroughly, and then he rolled gleefully over and over in the dust. The other little fellow all the time looked on with amusing curiosity. Finally he came up and pulled the bird skin away from his playmate. Thus the fun went on. The vixen, or mother fox, was probably inside the den, for it was not yet time for her to go out to hunt; but since it was fully my intention to come back and see more at another time I did not think it wise to disturb the family to find out. So when I saw both of the little foxes go into the hole, I slipped away.

I told an old prospector friend of my find. He surprised me the next morning by bringing in one of the babies for a camp pet. "Got it in a box trap," he said, "and almost had both of them." The little fellow soon overcame his fears, and proved unusually playful and created a great deal of amusement among the men at our camp. So mischievous was he that nothing within reach was safe from mauling. Old shoes, the corners of

THE GRAY FOX

blankets that hung down from the beds, sticks of wood from the wood box, everything had to be pulled and chewed by the baby jaws. As it grew older, the animal

Photo by Donald R. Dickey

ARIZONA GRAY FOX

wandered from time to time out into the brush to be gone for several days at a time, but always it returned and manifested the greatest satisfaction at being back again.

DENIZENS OF THE MOUNTAINS

The west coast of the United States is a fox paradise. No less than eight different races of the gray fox are found in California. Each of the coastal islands of southern California, gives a home to a distinct race. The Lower and Upper Sonoran zones of the southeastern deserts are inhabited by the Arizona gray fox. The humid coast belt of central California claims its redwood fox; while the region included between the interior of Humbolt County and Mount Shasta is inhabited by the Townsend gray fox. It is a curious fact that the fox found in the famous Rancha La Brea fossil beds of Los Angeles is apparently identical with these foxes now living in the far western mountains and valleys. "This probably represents one of the few if not the only carnivorous mammal," said Frank S. Daggett, "that passed unaltered in form through conditions that brought extinction or great modification to the fauna of southern California generally."

The western gray foxes exhibit little of the cunning and crafty intelligence shown by the red foxes of Europe and the eastern United States. Indeed they are so persistently and unsuspiciously foolish that they seem to be the dunces par excellence of the wild, and trappers find them to be among the easiest of our larger mammals to catch. They seem never to have been intended by Nature to live in a man-infested land. Neither their wits nor their cleverness is of the kind to help them out in the presence of man's trapping devices.

THE GRAY FOX

When left alone the gray fox manages to get along very well. He is a rustler in the matter of supplying himself with food. He never lies around waiting for something to turn up, but goes out on the hunt as soon as the long shadows of evening have crept over the mountain. On cloudy days I have seen him out hunting, but I am inclined to think that this is not usual. The fox comes to his meal at no spread table as does man; almost all of his food is alive, on the move, and ready to speed away from him at his approach. Often he must, because of the adeptness in flight of his intended prey, find himself without a meal just at that moment when he most fully expects to have one and when he is most hungry. We hope the alarming shrinkage of his menu is not taken too much to heart!

Competition between foxes and other animals for food must in times of fair abundance be small. Foxes subsist largely on rodents of which the supply is generally plentiful, small birds which roost near the ground, and insects. Indeed, much of the food of the fox is such as is generally ignored by the larger Canidae. In relation to the destruction of rodents, the fox's economic record is good, for he is a tireless destroyer of mice, ground squirrels, pack rats, and gophers. To sick and crippled birds, he proves himself a blessing, for he gobbles them up and puts a quick end to their misery. To the happiness of the strong and healthy ones he is a constant menace. This carnivorous night

DENIZENS OF THE MOUNTAINS

prowler, with his sharp eyes, keen ears, and quick snapping jaws, makes sad end of many a bird, her eggs and young. It is an easy trick for him to take the birds that nest or roost near the ground or in low bushes. A fox can climb into a small bush with ease, and even up onto the branches of a good-sized tree if the trunk leans at a fair angle away from the vertical. This animal is so sly, secretive, and owl-like in his silence of movement that roosting birds have little chance to evade him. We can not blame him if he eats the birds, for he was fashioned by nature to engage in such murderous enterprises. This does not alter our decided opinion, however, that his numbers should always be kept down by judicious trapping.

It seems to be good form in foxdom that the female should have entire charge of the rearing of the young. For her pups, she must do all the hunting, and in time of need it is she that is called upon to defend her offspring. The female, generally shy, shows absolute self-abnegation after the appearance of the young, and offers a most threatening front to any creature that disturbs the peace of her family. The number of pups is three to five; they are born in April, May, and June.

The gray fox shows no unusual ability as a songster. Perhaps he comforts and excuses himself by saying that he saves himself the cursing bestowed on the noisy coyotes, or that short songs are best after all for small folks who need to keep silence that their presence

THE GRAY FOX

be not known. A fox can hardly be said to bark in the true sense of that word, for his voice is little more than a coarse croaking yelp, given, not in series like the coyote's "song," but singly and infrequently. I hear it most often at dusk, when the foxes are just beginning to come out for the night's prowl.

The natural enemies other than man are few. The meat of canines, like the meat of birds of prey, is generally distasteful to other carnivores. I have known prospectors, who had dogs, to attempt to save on other rations by cooking up fox and coyote meat and then trying to induce the dogs to eat it. Though they had attempted to disguise the flavor by seasoning with onions, pepper and vinegar, their beasts would have none of it.

The gray fox is a creature of unquestionable beauty. What wonderful, brilliant, sharp, piercing eyes he has! As he looks at you he seems to "look you through and through," and you are almost able to feel a ponderable material, something going out from his animated eyes and piercing to the marrow of your bones.

CHAPTER XIV

THE WATER OUZEL OR DIPPER

IN 1915 the Colorado Desert was visited by unusually heavy winter rains. One of the storms of January seemed of the nature of an extended cloudburst, and the oldest of the Cahuilla Indians said that they had never seen a winter storm of equal magnitude since they were children. Water poured in such quantities from the heavens that the broad, shallow stream-beds of the open desert were turned into swollen rivers, and the mountain canyons bellowed and roared with the plunging streams of turbid waters that rushed along their bottoms.

Standing near my little shanty at Palm Springs, I could look up onto the steep mountain behind me and count at one time twenty-six waterfalls in the throats of the steep-walled precipitous gorges. Everything that would float or that could be moved along by water seemed about to go. Shrubs were loosed from their moorings along with rocks and sand, and the streams were glutted with debris from the highlands. It was a glorious and exhilarating sight to witness Nature in so uproarious a mood, and to see her playing with her forces in such abandon. Partially protected by a canvas thrown over my shoulders, I walked about in

THE HOME OF THE OUZEL

the storm for hours watching the Storm King at work and listening to the deep sounds of his voice. Rain torrents were falling all about me. Dark clouds hung low on the brow of the mountain. The whole of the broad desert valley was bathed in deep colors of blue, smoky gray and leaden lights. It was a picture of gloom but of marvelous sublimity.

In the midst of all the uproar of torrential flood waters, I heard ringing out loud and clear and strong the rich melodious song of a happy bird. There almost before me on a rock in the midst of the turbulent stream was a plump, short-tailed, slate-colored water sprite, the dipper, that captivating songster of mountain cascades, singing and dipping and rejoicing in the storm elements that sent all other birds to seek silence and shelter. More surprised I could not have been, for I had always associated this bird with clear mountain streams. To see him here in a muddy, intermittent, desert stream seemed indeed an anomaly. Up and down the foaming rapids he went, seemingly oblivious to all that was going on about him, singing in the gloom, and happy so long as he could have waterfalls and torrents in which to play and feed.

Storm days had been moving days for him, for this bird who came with the tempest now took up his home along the little irrigating ditch, the waters of which come so merrily from the mountains past my desert shanty. And there for many days I heard his rich

THE WATER OUZEL OR DIPPER

song poured out to the soul of the desert, or saw him fishing and diving and rollicking in the clear laughing waters. But the desert was good enough for him only in winter; he considered it no place for him in summer, and when the hot days of late spring came again he betook himself to the mountain gorges.

It is interesting to speculate as to the manner by which ouzels came to occupy such isolated, desert-bordered mountains as the San Jacintos of southern California. Are they descendents of birds that occupied this same region when it was yet directly connected with other high mountain masses? Or have they, or their immediate ancestors, migrated across arid land reaches from neighboring forested highlands? Other birds, such as the jays and mountain quail, have been seen crossing the San Bernardino Mountains to the San Jacintos, and it is not beyond probability that the ouzel, though much restricted to riparian environments, might find ways infrequently offered to make its way from one range to another. Does not this experience with the dipper in the storm suggest a possible explanation of the manner and means of migration? It seems entirely plausible to believe that an ouzel, following along the courses of intermittent streams, could, during heavy storms, cross the narrow arid strip lying in the pass between San Jacinto and San Gogornio Peaks.

Though it may not be generally known, water ouzels inhabit many of the mountain canyons contiguous to the desert, both in southern and middle California. They have been seen in Palm, Tachquitz, Murray and Chino canyons on the edge of the Colorado Desert, and along many of the streams flowing into the arid Owens Valley. Their essential environmental element is cold, rapid, running water and wherever it is found from Alaska to Central America, and from the Rockies to the Pacific, this water-loving bird is at home.

The ouzel is a bird that attracts attention to himself through his extreme individuality. He is in no sense a bird of the common lot but a creature of queer combinations, of unique habits and peculiar abode. Though in every sense a passerine bird, he has many of the ways of the sandpiper and the duck. Special adaptations have made it possible for him to live advantageously along streams, and for a great part of the time he is actually in the water. His feathered coat is waterproof because of its compactness and because of the oil that is on it. When the ouzel goes into the water his feathers hold to themselves bubbles and films of air, as do the chitinous coats of certain aquatic insects; consequently, when the bird emerges from the water his feathered jacket is as dry as before he went into the brook. The wings are not only efficient organs of flight in the air but also are a good means of locomotion in the water, and it is one of the queerest of sights to behold an ouzel making his way through the water on the wing. He is no web-footed bird but he gets along in the water

THE WATER OUZEL OR DIPPER

as well as many of the birds that have membranes between their toes.

The American ouzel is a bird whose affinities were soon recognized by the early explorers, who doubtless were already familiar with the European dipper (*Cinclus aquaticus*) of their own homes. They gave it the name ouzel, which they were in the habit of applying to the European bird, and this name it has retained ever since. The name "dipper" was apparently invented and bestowed upon the bird by Bewick, who in writing his "British Birds," desired to coin a name that would suggest the bird's habit of bobbing its body "in a dipping motion or short courtesy, oft repeated." Ousel or ouzel is a word of Middle English and Anglo-Saxon origin, which, at least in England, is also applied to a blackbird, the king ouzel.

While the adventuresome mountaineer or fisherman threads his way up deep, damp, sunless gorges of the Sierras, the dipper is often the only bird he meets and his song the only voice of melody that greets his ear. This little stubby-tailed and plainly attired vocal hermit delights in following the hurried streams even though his journey takes him into dark and forbidding places. So "wedded is he to wetness" that you seldom see him unless he is actually in the water or sitting on a rock amidstream or by the water's edge. While he is always cautious, he seldom resents your approach, and it is possible by slow movements to slip very close upon him—so close that you may easily watch almost every movement as he drops his courtesies or dives and swims in the running brook.

The food of the ouzel is aquatic insects and their larvae, also snails. It is this that accounts for many of the frequent dips he takes into the water. When he goes in after his food he does not dive off the rock but jumps or deliberately wades right in and then makes his way beneath the water until he finds what he wants. The more the waters boil and foam the better he seems to be pleased. Now you see him sitting, bobbing up and down on a mossy rock, then in an instant he is wading into the water, ducking his head and then going altogether out of sight, only in a very few seconds to bob up again and repeat his funny motions.

Up and down the stream the ouzel works his way, fishing in almost every pool and rapid near at hand, but if for any reason he desires to go to new feeding waters he finds himself an able flier. The flight is a very direct one and fairly rapid. The wing beats are very steady and the bird utters strange, "clucking notes" as he speeds along. When the dipper alights it is almost always on a rock or log. If he seeks rest on a tree it is on its kneed roots on the edge of the stream bank and not up in its branches. If you frighten an ouzel, there are but two avenues along which he will fly, upstream or down. Unless the occasion should be an extraordinary one he would no more think of flying

THE WATER OUZEL OR DIPPER

out into the brush at the streamside than he would of flying upward like a skylark.

The nest of this peculiar, aquatic passerine bird is framed of green mosses, sticks, leaves and grasses,

NEST OF THE OUZEL

built up in the shape of a low, squat oven with a somewhat depressed opening at one side for entrance and exit. Many times when the approximate situation of

DENIZENS OF THE MOUNTAINS

the nest is located it is not easy to put hands on it for it may be built in a place so difficult of access that the observer fails ever to see it. It is generally placed on some ledge or rock shelf underneath a waterfall or beside a streamlet where the young are well protected from almost all of those enemies common to other birds. Concealment of the nest is aided by the bird's habit of adding to its exterior moss which stays green in the damp nook where the nest is placed.

"However sly and shy," says Dennis Gale, "this bird may be, when it is satisfied that you have discovered its nest all shyness and slyness end and then a more confiding fearless little fellow is not to be met with. Going in and out of its nest even when you have your hands upon it, the bird with plaintive appeal both in speech and actions seeks to gain your sympathy. Rob her of her treasures and she will rebuild and lay a second clutch in three weeks, and that taken a third will engage her cares. This fruitful industry no doubt is often exercised independently of the interference of man for with the sudden rise of streams the nests must often be swept away. The nest is seldom betrayed by the bird itself who will slip from it unperceived, even while the observer is looking at the spot, and drop into the water like a stone, then float or swim under water several yards down the current before coming to the surface. It manages its flight to suit circumstances. While interference with its nest is threatened it will

THE WATER OUZEL OR DIPPER

fly about with the address and lightness of a hummingbird, noiseless as an owl; upon other occasions it will dart down the creek with lightning rapidity." (Unpublished manuscript, University of Colorado Library.)

CHAPTER XV

THE GOLDEN-MANTLED GROUND SQUIRREL

WHEN I begin mentioning to mountain folk my interest in chipmunks and their ways, I am so often met with the remark, "Now there are two kinds here, little lively ones that go up trees, and big yellow-headed ones that stay on the ground; and then on the desert there is yet another kind with a white tail." And when I attempt to convince them that neither the large fellows nor the white-tailed ones they mention are chipmunks at all, but small ground squirrels, they put on an air of wisdom and shortly inform me that they know better, for these creatures they have in mind "have stripes on their sides, and the big ones of the mountains play around with the little tree-dwelling chipmunks." Thus is registered again a widespread error. And then the professor must begin once more to explain that the difference between a chipmunk and a ground squirrel is not one of size but of anatomy and coloration; that a ground squirrel's stripes run only up to his shoulders, whereas those of a chipmunk run right up to the end of the nose.

I will admit that both the golden-mantled ground squirrels of the mountains as well as the little antelope ground squirrels of the desert, which unfortunately are

THE GOLDEN-MANTLED SQUIRREL

called chipmunks, do possess remarkably small chipmunklike bodies and have many near-chipmunk ways. If we give careful attention at all to their habits, however, we soon see that the seeming likeness is more apparent than real.

Courtesey U. S. Biological Survey.
THE GOLDEN-MANTLE OR CALICO SQUIRREL

Golden-mantled ground squirrels are burrowing animals of rather local distribution in California, being seen for the most part only in the higher parts of the San Bernardino Mountains and Sierra Nevadas. Beyond the borders of the state they are to be found in the pine forests of many of the western mountains, including the

DENIZENS OF THE MOUNTAINS

Rockies from northern British Columbia to New Mexico and Chihuahua in Mexico. The author has studied especially the habits and mannerisms of the subspecies inhabiting the mountains of southern California and the High Sierras, and to these he will for the most part confine his remarks.

In the region just referred to the "copper heads" as they are sometimes called, begin to make their appearance as we come into the forest of yellow pines. Above this they range upward through the belt of firs and into the region of lodgepole pines. Nowhere do they occupy dense forested areas, but make their short burrows under rocks and fallen logs in open spaces where the sun has free access. They are very partial to sunshine and spend many a happy hour basking upon the rocks and exposed stumps.

Often I have been sauntering through the high mountain forests quite unconscious of having any animal neighbors about me and have realized afterwards that I had dozens of little ground-squirrel eyes watching me all the time. For as soon as I sat down and was quiet for a few minutes I saw, now here now there, one after another, these little calico squirrels mounting the rocks and tree stumps to look at me. When I was making camp near their colonies, they have always kept a close eye on me. At first they seem to be timid creatures and scamper away into the brush or under logs whenever I advance toward them. But as the days go

THE GOLDEN-MANTLED SQUIRREL

by, they become bolder and bolder and, though still reasonably cautious, go about their business much as though I were not about.

In the summer of 1925 I went into camp a number of days at Palisade Creek. At this place I was much impressed with the extraordinary boldness of these squirrels when once they found it safe to come in among the pack boxes. Though repeatedly chased away they came upon the tables, moved lids off the cooking vessels, and made themselves a general nuisance. When chased off they ran a short distance out into the open, sat up and watched us and then, the minute we were quiet, regained their former position and were into the pots and pans again. Sometimes two would coöperate in raiding the cupboard. One would sit up at a distance and watch, ready to give a sharp alarm note if he saw the other being approached by one of us.

Their omnivorous appetite manifested itself in the eating of almost everything within reach; meat scraps, bread, dried fruits, potatoes and pastries, were all consumed with relish. Once I saw two of them fall to eating a hot pumpkin pie. So warm it was that the cook felt for once that he had something they would not touch. They first cautiously nibbled around the edge, then finally, after they had finished most of the crust, boldly jumped right into the middle, now almost cold, and ate greedily there. Though the filling was so

THE WESTERN PORCUPINE

THE GOLDEN-MANTLED SQUIRREL

delicious, they knew better than to stuff their capacious cheek pouches with it. I suppose they fully realized the folly of putting into these furry pockets things that they could not take out readily and with ease.

One day I threw out the wrapping paper from several pounds of butter we had just unpacked. One of the pieces was about nine inches square and considerably wrinkled. Hardly was it on the ground before a golden mantle was after it. She seized it at one corner in such a way that it was difficult to see over its edges, and part of the wrinkled paper trailed at her side. In her flight she ran up onto a prostrate log and plump into a small chipmunk. The little fellow was so taken by surprise that he almost fell over backwards before he could collect himself and sputter his curses on the inconsiderate lady. The golden mantle, undisturbed in her impoliteness, gathered the paper up into a ball with her fore paws and then took it away in her mouth. Jumping from the log, she skimmed along the ground and went into a hole under a rock. I quickly secured a shovel and crow bar and tried to dig her out, with the idea of finding out what kind of a place she had underground for her living quarters. There were four openings near and these I carefully plugged with little stones. The burrow I was working on first went horizontally in under the rock about six inches, then downward and inward again. The soil was so very dry and powdery that I soon lost the direction. I then tried by digging all around to find the tunnel again but my work was in vain and I never did find my squirrel. She had prepared her home against just such inquisitors as I. I hope she found the buttered paper to her liking, and while eating it enjoyed the hidden laugh on me.

Let us now show up the cruel pugnacity of the golden mantle. I can state from my own experience in handling them that they are frightful biters and can inflict severe wounds. An artist, whose word we may trust implicitly, tells me that yesterday, while he was out sketching, his attention was drawn to a calico squirrel rapidly pursuing another of its kind about a rock. The first one had chased the other in hot pursuit but a few yards when they began what appeared to be a "wheel motion," the one going over the other in such harum-scarum fashion that it was difficult for the observer to follow their movements. Finally the agressor, like a weasel, snapped his powerful chisellike incisors into the other's neck and began dragging him triumphantly off through the bushes. The poor creature now limp and helpless was wholly at the mercy of his captor. My friend would have liked to have followed them and to have seen the end of the struggle, but his heart so went out for the poor victim that he frightened the agressor off. The wounded squirrel was left much weakened, but after a few minutes it somewhat revived and made its way slowly off the scene.

These two squirrels were of about equal size, but the

THE GOLDEN-MANTLED SQUIRREL

one that carried off the other did it with little effort. The whole affair was over in perhaps three-quarters of a minute. There was considerable noise about it. Gutteral, squeaky notes were uttered by both animals throughout the contest.

Knowing what I do concerning the meat-eating propensities of these animals, I venture to say that had the one ground squirrel killed the other he would have eaten its flesh. It is not above them to eat carrion.

Golden mantles do not object to wetting their jackets and are able to swim efficiently when put to their wits in the water. At the junction of the Middle Fork of the Kings River and Palisade Creek, one was thrown into the water and not for a moment was he at a loss to know what to do, even though the water was here very swift. Two strokes carried him outward toward a little island. Reaching it, he immediately turned and made for the shore, attaining it with apparent ease and greatest celerity.

Food competition between these squirrels and the chipmunks is limited to those foods that are sought on the ground and low bushes. Were not this the case the golden mantles with their agressive ways would doubtless displace the chipmunks in any area in which they came to live. The chipmunk has the saving advantage of being able to get some of its food in the trees. Pine nuts, and other tree seeds, bark insects, and fungi are food resources always within its reach, but largely

DENIZENS OF THE MOUNTAINS

denied the golden mantles because of their preferred terrestrial habits. I have seldom seen a golden mantle go more than four feet into a tree or shrub, and never more than eight feet. Their clumsily built bodies disqualify them as skillful tree climbers. At Grouse Meadows in the Sierras I saw one go up into a willow to get some cheese rinds which were lodged in a crotch. He was so awkward that he soon lost his hold and would have fallen had he not caught himself and wheeled around the branch to a safe position.

Comparatively early in the autumn (generally in late September) the golden mantles seek hibernation quarters and then lie in somnolence until spring sunshine brings sufficient warmth to revive them from their stupor. In the San Bernardino Mountains I have seen them out gathering the seeds of the lead plant (*Ammorpha californica*) as late as the first of November. It is probable that the southern California "callos" emerge from their hibernation correspondingly early in the spring.

The young number from two to six, and are generally born in July though late June and early August records are known.

CHAPTER XVI

THE WREN-TIT

ANYONE with a desire to become acquainted with this brush-loving bird may easily come to know him, for he advertises himself along almost every chaparral-bordered foothill road by his series of loud staccato notes and trill, which he gives with such remarkable frequency, *keep-keep-keep-keep, keepit-keepit, keepit.* The intervals between the notes are such that one can not but think of a bouncing ball. And as though the singer meant that you should not miss him because of any silence in song, he keeps to his resonant, ringing, trilling notes throughout the year, excepting possibly a brief season of rest during the moulting time.

Since the wren-tit is a low-flying bird he is easily seen. His almost uniformly smoky brown, fluffy-feathered body, his unusually long tail, and jerky flights from bush to bush are distinguishing field characters of the most helpful kind. If you are patient, and will secrete yourself or even stand still in the scrub forest of his home, and will imitate his song, by making a squeaking noise, his jaylike curiosity will undoubtedly impel him to hunt you out. He wants to see you and will come within a few feet of you. As if fascinated by your

Photo by Donald R. Dickey
PALLID WREN-TIT

THE WREN-TIT

presence, he will for moments hop inquisitively about on the ground or from shrub to shrub, eagerly peering at you from every angle. The chances are that all the while he will be giving his gentle peeping note and making you feel that he is inviting you to respond to his call. Could any bird give you a better opportunity to know him? Those beginning to study ornithology generally complain that the greatest impediment to their progress is the shyness and restlessness of the birds. Indeed, most birds are so discouragingly restless that hardly are they located before they are off on the wing and out of view. But here is a bird that, like the rock wren, comes to you if you will but have patience in waiting, that one thing you must possess if you would know your bird neighbors. The wren-tits often forage in pairs and it is quite possible that you will find both birds coming near. But you may have difficulty in recognizing the identity of the sexes for they are indistinguishable even in the hand. Only by means of a dissection is it possible to separate them.

Any attempt you may make to study a wren-tit at close range by trying to steal up to him will surely end in failure. He moves as fast as you do and is always an adept in slipping out of sight just at that moment when you think you have him in a place convenient for observation.

The nests of wren-tits are exceedingly difficult to find. Again and again I have watched the adult birds

DENIZENS OF THE MOUNTAINS

and have known from their actions that nest building was going on or that young were in the nest, but the number of times I have been able to find the domiciles is few. The parent birds are very close sitters and are flushed from the nest only when the intruder comes right upon them. Then instead of making show of excitement and attempting to lead the unwelcome guest away by feigning lameness or broken wing, they escape as quietly as possible.

Observation of wren-tits during the nesting season leads me to believe that the time which elapses between the day when the first eggs are laid until the young come forth is about four weeks. Within a little more than two weeks after the birdlings break the shells they are able to leave the nest. They are still for some days under the protection of the parents, however, and upon the latter they must depend for a major portion of their food, being too half-witted and inexperienced to feed themselves. It is always amusing to watch the awkward, fuzzy-feathered youngsters as they sit on twigs with wings all aquiver and their mouths agape while they impatiently chirp their baby songs and express their eagerness for the caterpillars the parents are constantly bringing in to them. The parents may seem to attempt to feed the young so that each will receive its due share, but many are the times that the most agressive birdling gets more than his quota. The weaker ones get what they can, but all too often

THE WREN-TIT

that is far too little and starvation eventually becomes their lot.

It is common knowledge that baby kittens are born blind and that they continue to be sightless for some days after birth, but it is not so well known that many wild birds come into the world with closed eyes. In the case of baby wren-tits, the eyes do not open until five or six days after the eggs hatch. Moreover, the young birds are at first well nigh naked, and so present a picture of utter, almost pitiable, helplessness. Nature anticipating their early need of clothing soon comes to the rescue, however, and by the end of the fourth or fifth day the tiny bird children are quite well clad with feathers.

The range of the pallid wren-tit in the United States includes most of the coastal and foothill region of California, from the vicinity of San Luis Obispo Bay south to the Mexican border. In the San Francisco Bay region lives the intermediate wren-tit, while in the Coast Range Mountains of Oregon and Washington is found the coast wren-tit, a variety with ruddy-brown underparts. In southern California this bird is for the most part confined to the brush-covered hill districts beyond the limit of towns. Its song is most frequently heard along the chaparral-bordered trails approaching the edge of the pine forests. In Berkeley I found the wren-tits very plentiful about the university grounds. In fact, during the spring days, the ringing scale-like trill is one of the most familiar bird songs found on the upper campus.

Like the California thrasher, and the familiar towhee or browny bird, the wren-tit is a non-migratory, home-loving bird, staying year after year in the vicinity where it first takes up its home.

The common English name for this bird would imply that it is a mixture of wren and titmouse. While it is now regarded as nearer to the wrens than to the titmice, it is given a position in classification not near to either. No definite niche could be found for this bird among the many other bird families and so it was placed in a family of its own and it remains today a monotypic bird, the only species of the only genus of the single family *Chamaeidae*, a distinction enjoyed by few birds.

CHAPTER XVII

THE WESTERN STRIPED SKUNK

WHILE I was camping in the Palomar Mountains one summer I made a journey of twelve miles thrice weekly to the post office at Nellie. The road is a tortuous one along the side of the mountain and in the whole distance there is but a single "cut off," and that a very short one through an open space in a forest glen. To save time, the little brown burro and I always went that way.

One day I noticed in the ground beside the trail a small hole, about the size of a lead pencil, and in and out of it were passing black-and-yellow banded wasps. Beneath was their underground home and often afterward I stopped to watch them. As though I was unworthy of their regard they seemed to pay no attention to me, and I came to look upon them as wholly unobjectionable creatures.

A big striped skunk passed on the short cut trail one night, noticed the wasp domicile even as I did, but saw in it what I did not—a place to get a meal. She lost no time digging into it, now well stocked with fat grubs, the pupating young. She dug down nearly a foot and a half, cleaning out as far as I could see, every one of the juicy white larvae. I don't know how many of the stupid old yellow jackets she killed in addition, but there were enough left the next morning, when I went by, to make a good lining over the bottom and sides of the hole. They were slowly crawling about over one another and in a mind as unstable as fermenting yeast. Like revengeful parents they were ready for any new developments.

Evidently they had decided to vent their rage on the very first moving things, great or small, that showed themselves near their despoiled and plundered domicile. And those first moving things happened to be a poor little innocent burro and myself. No sooner did we come into their presence than up they flew, the whole swarm of them, and they went after us in pell mell fashion, digging into the hair of the burro and stinging her with unmerciful vigor. I received my share of waspish wounds behind the ears, and on the nose and neck, and before I was able to do more than take in the situation the terrified donkey had struck up the hill on a bucking gallop that sent nearly all of the packages in her pack rolling to the ground. I have reason to believe that those hornets were satisfied with the outcome, for they had done well their part in dismaying the enemy. The little burro never forgot her experience on the short cut trail and during all the rest of the summer I could never get her to pass that way again. Once had been enough. Burros remember. I still have a settlement to make with that skunk.

THE WESTERN STRIPED SKUNK

Several species and subspecies of striped skunks are known to inhabit the mountainous districts of the western United States. They are much larger than the little spotted skunks and, though the amount of black and white varies, all of them carry a white stripe, which, beginning at the nape of the neck, divides on the shoulders and then runs along the upper parts of the sides and across the hips, ending on the sides of the tail a short distance from its root. When seen with their plume-like tail carried well above the back, skunks are animals of extraordinary beauty, and this fine appearance compensates somewhat for their other objectionable propensities. The appearance of no forest animal is to me more attractive, and I know of few prettier animal pictures than a mother skunk and her kittens moving along in a forest flooded with moonlight. The attentions to one another are most gentle and intimate. When the mother goes digging like a bear in the leaves beside a log for fat grubs, there is only the most eager and playful rivalry for the juicy prizes, never the manifestation of jealousy and never a quarrel.

The skunk is guilty of some sins to be sure, but it also possesses some good qualities and there is much to be said in its favor. Skunks destroy enormous numbers of harmful rodents, and as consumers of noxious insects deserve special mention. Their diet is a varied one and we are sorry to have to record against them that they sometimes eat bees and small birds and rob

DENIZENS OF THE MOUNTAINS

the larger species of their eggs. It is well to remember, however, that these constitute but a minor portion of their food.

One can not long sleep out in the open at night without coming into close relations with the inquisitive skunks. Only last night I was twice awakened by hearing them scratching in the oak leaves beside my bed. On another occasion I awoke with a skunk on the pillow beside my head. When I raised myself that I might look my intruder square in the face she fled from my presence more frightened than I was. As I saw her moving away, I thought my experience with this skunk was at an end. But not more than two hours afterward I saw her emerge from an old hollow log, and with her came five little replicas not over several weeks old. They all came trooping along after her in single file after the manner of ducklings and looked on with greatest interest when she sank her nose into the leaves and scratched for insects. It was amusing later to see her sniffing about the pack boxes at the head of my bed while all the little ones walked over my covers with never a manifestation of fear. I have often wondered since if they were even conscious that I was there watching them. The ways of this family were certainly civil and even neighborly, and I have yet to regret their familiarity.

The spring preceding I had purchased a young fawn-colored burro. Life was yet very interesting to her and

THE WESTERN STRIPED SKUNK

on every turn new things presented themselves to her for investigation. She seemed to rely on her nose more than on any other sense organ to inform her of the nature of anything strange. A constant stream of messages seemed to be pouring in through it. Her first camp fire was investigated with her nose, a course which I assure you brought its own obvious punishment. The first stream she came to was not crossed until she had thoroughly smelled it, water and banks.

On this particular evening my pack train was slowly moving along the trail when a skunk unhesitatingly emerged from the brush at the trailside not more than ten feet in front of this inquisitive burro. Like everything else that was new, this strange moving thing must not only be carefully scrutinized with the eye but immediately investigated with the nose. My burro now, in spite of all my efforts to hold her back, began trotting after that skunk, gaining on the deliberate creature all the time. I knew what would surely happen if she ever touched it. But just as she was about to come to triumph in naughtiness and I was thinking of what dreadful odors would be mine to endure, the skunk redeemed all the debts of her tribe by slipping beyond reach into the brush and leaving the young donkey looking foolish and disappointed, like a hound that has lost its quarry.

The striped skunk is typically an animal of the low mountains and bordering foothills, and chooses for its

DENIZENS OF THE MOUNTAINS

home, in so far as possible, wooded districts close to streams or other available supplies of water. The feeding range is not extensive, a half-mile being about as far as they ever go from the home burrow. This statement does not apply, however, to the males during the breeding season. Urged then by the strong desire of sex they wander far and wide through the forest. Recently I happened upon the tracks of a stump-footed skunk and following it I came upon a place where two skunks had quarreled and the fight that followed was evidenced by blood upon the leaves. Several nights afterwards I caught this same stump-footed skunk in a trap about five miles up the stream. It was a male and he was minus an ear. The root of it was still raw flesh. Evidently it had been lost in the fight I have just mentioned.

Skunks are not good diggers in the same sense that burrowing rodents are, but they will excavate a burrow in which to deposit the young if one can not be found already existing in rock crevices or under logs. The kittens are borne in April, May, or June, and number from five to nine. They remain with the mother the first season and may "hole up" with her the first winter. By the time spring arrives again they have sought shelters of their own and are hunting for themselves. Claude T. Barnes in his "Mammals of Utah" tells us that the young are able to emit the strong odors when one month old.

THE WESTERN STRIPED SKUNK

If it is ever desirable to catch a skunk it may easily be taken in a box trap and killed without odor by drowning. Of all the instruments of attack there is no more foolish one to use than a pitchfork, yet how frequently is this weapon employed. The animal is certain to emit the fetid secretions when so horribly killed and there will be a reminder of the struggle for many days to come. The odor of the striped skunk is much more lasting than that of the little desert "civet." A little of the oil may contaminate a large volume of air or water. As evidence read the following which appeared as a news item in a recent daily of a southern California city.

**WATER IS OKEH AGAIN.
HERE IS HOW IT HAPPENED.**

Riverside folk need no longer complain of the domestic water of the city which caused much uneasiness yesterday among the many users because of the bad odor which emanated from the water. The city health physician announced last evening that the source of the pollution has been discovered and steps were taken at once to remove the cause of the trouble which was at the Mt. Rubidoux reservoir.

Careful investigation made after complaints were sent in from all sections of the city revealed that a skunk had fallen into the reservoir in some unexplained manner, notwithstanding that the large tank is screened tightly.

DENIZENS OF THE MOUNTAINS

The reservoir has been washed three times with strong solutions of chloride of lime. Every inch of the interior was scrubbed thoroughly.

Following first reports of the condition of the water all the reservoirs in the city were emptied. First efforts failed to disclose the source of the annoyance. All residents were then requested to use as much of the water as possible for the sprinkling of lawns or gardens, with expectation that the water would soon be cleared. Fire hydrants of the city were also opened, but the annoying conditions still existed. Another inspection of the Mt. Rubidoux reservoir revealed the trouble.

It is estimated that fully five hundred thousand gallongs of water were contaminated by a fraction of an ounce of skunk scent.

In spite of its bad odor the striped skunk is not immune from attack by other animals. Coyotes have been known to attack them and Dr. J. Grinnell and H. S. Swarth report finding a portion of a carcass of one in the nest of the golden eagle. The nest contained a single young bird yet too small to fly and the skunk had evidently been brought into it as a part of the food supply. There is evidence to support the belief that in these birds the sense of smell is vestigial or entirely wanting, and so we are led to believe that this foul-scented animal was fully acceptable to eaglet taste.

#201 02-25-2006 9:58AM Item(s) checked out to WILD, PETER T.

TITLE   BARCODE
 DUE DATE

Our desert neighbors.   39001007185864
 08-24-06

Denizens of the mountains 39001007185831
 02-26-06

            Renew your books online at http://sabio.
                    library.arizona.
edu/patroninfo/

CHAPTER XVIII

THE JUNCO OR SNOWBIRD

WHO is this lovable, jaunty, little mountain bird, that comes to us dressed in brown with sooty black head and shoulders, and with white outer tail feathers that show up so prominently when he flies? Though he is such an intimate dweller about our mountain camps during summer, and so sociable, fearless, and confiding that everybody feels acquainted with him, yet few can tell his name, though they ever so much would like to know. Well, he is the Sierra junco (*Junco oreganus thurberi*) or, if you like to give him a more simple name, you may call him the "tick bird," for his ordinary call note which he is constantly giving is an explosive *tick-tick-tick*. But do not think this is his only song. Many an evening during the nesting season I have seen him sitting on a pine stump and heard him making as merry a warble as one would wish to hear. The only disappointing feature about the song is that it often ends so abruptly as to seem incomplete. You are always waiting to hear the finish and it never comes. The female junco is a quiet little lady dressed in quaker gray, and her note is a very modest one.

These black-cowled juncos, or snowbirds as they are

Photo by Donald R. Dicky

THURBER JUNCO—NEAR NEST

THE JUNCO OR SNOWBIRD

also called, are common all summer in the yellow pine belt of our mountains, even ranging to the pinnacles of the highest mountain peaks. I have repeatedly seen them on my visits to the summits of San Jacinto and San Gorgonio peaks, and on the high crest of the Sierras. During the summer season, they are usually seen alone, in pairs, or a few together, but in the autumn they consort in flocks varying from a few birds to several hundred, and they then wander through the forests together, hunting seeds and picking up the few insect stragglers that then may be about. During the heavy mountain storms of winter, the juncos come in great numbers down into the desert and coastal valleys, and there you will find them busily feeding on the weed seeds in the fields, along roadsides and even in vacant lots about the villages. But they remain only a few days and are generally back to the mountains upon the first appearance of sunshine and moderating weather.

The juncos build their nests flat on the ground, generally quite near streams or in fairly damp meadows, the very places where grazing cattle can trample them or where the water snakes are likely to find both eggs and young. Only this morning some little children brought me word that a water snake had taken the baby juncos from the nest below their house near the stream. The children discovered the frightened and bewildered mother bird in a high state of excitement flying above the nest, chippering and screeching in her

DENIZENS OF THE MOUNTAINS

inability to frighten off the murderous serpent, which was swallowing her young alive. With the help of their

Photo by L. M. Huey
NEST OF THURBER JUNCO

mother the children had killed the snake, and found on looking into the nest that already all but one of the

THE JUNCO OR SNOWBIRD

baby birds had been taken. The grief-stricken bird parent was so stunned and stupified with fear that for a whole hour she sat quiet and almost motionless above the nest, afraid, doubtless, to approach it, not knowing what further dangers might lurk there. Not even the squeaking cries of her last lonely, hungry birdling could arouse the mother instinct sufficiently to overcome the emotion of fear.

Water snakes, though possessing but feeble sense of smell, easily find the eggs and young birds and destroy many of them every year. Since they also devour great numbers of trout in the streams, I am inclined to question the wisdom of trying to protect them. While they doubtless prey, to some extent, on insects, it would appear that this virtue is far offset by their habit of eating fish and the eggs and young of insect-destroying birds. Better it is to have dozens of beautiful, beneficial birds alive than one wriggling snake.

When nest building is entered upon in the vicinity of mountain camps, the juncos are by no means bashful birds. Indeed they are a little indiscreet in choosing the home site. One nest I found being built almost within hand's reach of my camp table, and a second was made under a few brakes within four feet of my bed. In spite of the fact that I was continually tramping about near the spot, the mother bird seemed quite satisfied with the situation of her nest. So cozily was it hidden under its canopy of fern leaves, and so pretty were the three, purple-speckled, bluish white eggs within it, that it was a great temptation to look at it many times a day. The building materials were hair and rootlets, all tucked away in a pocket of pine needles.

Some days after the discovery of the nest, just after sundown I heard a peculiar hissing sound. Suspecting that there was a rattlesnake about, I approached the nest only to find that the strange noise was made by the birdlings as the mother brought in food for them.

The alarm note of the junco is a very rapid ticking sound. At Flower Lake, on the east side of the High Sierras, I noticed early one morning a strange commotion among a number of birds down near a little stream that runs from the lake. A strange medley of bird noises came from half a dozen white-crowned sparrows and a junco. The little junco was "ticking" so fast that it seemed like the sound of an over-hurried pendulumless clock. The birds jumped about so excitedly among the twigs that I knew someone was up to mischief. The birds are always first to spy out animal thieves and depredators. Sure enough, as I pushed my way through the brush, there ran out a small weasel, bent, I judge, on nest robbing.

The juncos are recognized by everyone as splendid little camp helpers, being diligent scratchers after cast-out crumbs. And there is no need of any one's being lonely so long as they are about. The more one gets acquainted with these little black-cowled birds,

THE JUNCO OR SNOWBIRD

the more the feeling grows that they want to confide and talk to you, and that their happy, snappy *tick-tick-ticks* mean something all the while. Not so with the intrusive, clamouring jay. You admire him in his rich coat of blue and cocked crest, but as you see him chasing companions with harsh-voiced outcries from tree to tree and garrulously wrangling with his mate you can have only a feeling of cold disgust toward him. For me juncos instead of jays every time.

CHAPTER XIX

THE MUTILLIDS OR COW-KILLERS

ALWAYS when the fine sunshiny days of autumn come, there arises within me the longing to get into the saddle and travel. Only those who have experienced the joy of sauntering or riding slowly over mountain trails can fully appreciate how deep is this desire to get out into the open to enjoy the genial friendly sunshine of the year's declining days. The fine spicy odors of frosted leaves coming on the cool breezes that blow from the throats of mountain canyons, the melancholy sounds of the crickets, and the return of the bluebirds to the valleys only serve additionally to remind one of how good for travel the mountain trails then are.

One day of a long-ago autumn I found myself instinctively, I may almost say, getting out the ropes and journeying toward the hill pastures to get my animals that I might make them ready for travel. And the next morning while the aroma of the greasewood and sumac was yet heavy on the air I was on my way, bound for the mountains.

As I rode along during the forenoon hours I came from time to time upon little parties of Indians. Some of them were camping along the roadside, while others

THE MUTILLIDS OR COW-KILLERS

were trudging along on ponies or in wagons toward the hill country where, they said, there was to be a fiesta at San Ignacio on the following Wednesday. These Indians were all as happy as jays and were dressed in those bright colors of calicos and woolens that western country stores are wont to supply. As they traveled, the old men sat on the wagon seats and the women and children on straw in the wagon beds, while the bigger boys and young men rode alongside on ponies. Never will I forget that brightly dressed, laughing-eyed little girl who chewed corn on the cob and kept smiling at me when from time to time I waved my hand at her as I rode along behind her father's wagon.

For making a journey interesting, there is nothing like having a definite, worthwhile destination. Up to this time I had had none in mind and I decided after talking to a number of the Indians that, inasmuch as I had never attended a primitive Indian festival, I, too, would go to the fiesta. When I announced my intention to my fellow travelers they all seemed much pleased.

As I neared the reservation at Cahuilla a splendid looking young Indian boy of about sixteen rode up and joined one of the companies which was traveling just ahead. Much to my pleasure, after greeting his friends, he dropped back and rode alongside of me. He had been away to school, he said, and wanted to talk to white people so he could learn to speak English better. His own people conversed only in Spanish and the native Cahuillan. This spirit of inquiry and desire to learn was rather unusual, and I could not but welcome his company and the opportunity to talk with him.

Realizing that these young Indian boys often have interesting experiences with the wild creatures of their mountain home, I was only too glad to engage him in conversation concerning natural history subjects. As I suspected, he was well versed in nature lore.

"See there," said Ortego, presently pointing to the ground. "Don't you see that thing there crawling in the dust? They are very bad things and you must never sleep on the ground if you ever see any of them around. They bite and sting, and if you are badly stung you must die."

It was a strange looking creature of the insect world that he pointed out to me. It appeared like a very large black-bodied ant, with abdomen well covered with a pile of deep reddish brown hair.

"What do you call him?" I asked.

"Some people call him a cow-killer and many hereabouts call him man-killer, and I think that is the best name."

"Did you ever see a white one that looks like a little wad of cotton crawling on the ground?" continued Ortego. "The white ones are males and the red ones like that one, my father says, are females. There are some man-killers too that have wings and some have

THE MUTILLIDS OR COW-KILLERS

short hair instead of long hair on their backs, but I don't think all of them sting and make you die, for once I was stung and it only made my arm hurt like fire for an hour or two."

The small insect he had pointed out to me was crawling very fast and nervously over the ground, and since I wanted to examine it more closely I dismounted from my horse and caught it in my handkerchief. This was an act which caused my young Indian friend no little anxiety and he assured me again and again that the sting might cause my death. This was my first real introduction to the queer little insect, for though I had seen the "cow-killers" about before they had aroused no special curiosity. Since that time I have never looked upon them without interest and it has been my pleasure to have them many times under close observation. The statements that Ortego made to me concerning them are, for the most part, not to be taken as fact. The beliefs that he entertained concerning their poisonous nature are, however, similar to those held by many rural people who see them.

Though the cow-killers or man-killers appear like ants they are in fact solitary wasps. They are plentiful over a great portion of the arid southwest and may be met within the mountain foothills, and up to at least an altitude of 5,000 feet in the mountains proper. They seem to avoid damp places and to spend most of their existence on the driest, hottest exposures. As

DENIZENS OF THE MOUNTAINS

they run hurriedly about and display their fine velvety coats in the sunshine, they appear to be such interesting creatures that there are few strollers who have not at some time or other been attracted by them.

If you are in doubt concerning the ways of the mutillids, as these insects are properly called, the books

THE MUTILLID

will not help much in removing your ignorance. Entomologists have never so far worked out the full life history of any of our western species, and what they have said about them may be found in but a few notes scattered here and there in scientific journals. An old but witful writer has somewhere remarked that the poets who lived before us have stolen all of our best

THE MUTILLIDS OR COW-KILLERS

thoughts so that little new remains to be said by the writer of the present. But this hardly holds true for anyone who would write of the mutillids. Some one who has it in his heart to do something original and worthy of commendation might well devote some time to observing these creatures and find out some definite things about the many obscure portions of their life history.

It is really remarkable how tenacious to life these hard-bodied insects are. If the soil is at all pulverized or sandy one may trample upon them with the full weight, and instead of their suffering fatal injury, as we would expect, they are but pressed into the dust and when released straighten out their legs and move off again apparently as spry as ever.

Because of their stinging proclivities, the Navajo sheep herders hold them in much dread. These pastoral people spend a great deal of their time lying about on the ground, even during the daytime, and they naturally come frequently in contact with such ground-dwelling insects. The mutillids are not aggressive with their weapons, as are the bees and paper wasps, but at times when they are ruthlessly trampled upon or otherwise molested they do not hesitate to bring the long needle-like stinger into use. The stinging organ is surely a formidable appearing weapon and if the insect were capable of thrusting it into our flesh to its full length the injury it might inflict would be considerable. I have

DENIZENS OF THE MOUNTAINS

seen a large mutillid unsheath a sting fully three-eights of an inch long. Only the females are armed with stinging organs. They may be further known by the absence of wings. The males are fully winged, and for some reason much less seen.

If you would know of the mutillid's ability as a "songster" and excavator take one up and drop it into a drinking glass into which you have previously placed a little sand. If the insect is now disturbed with a stick it will set up a high-pitched squeaky humming note, resembling one that can be made by rubbing a wet finger about the rim of a wine glass. You will notice that as the fine, birdlike, almost musical note is made black rings appear at about the middle of the furry abdomen. These are produced by the parting of the hairs and the exposure of the black exoskeleton when the rear half of the abdomen is rapidly vibrated. I am quite inclined to think, though I am not yet certain, that the note is made by forcing air out through the tubular breathing tubes or spiracles. The black antennae are very often rapidly vibrated in unison with the abdomen. As the mutillids crawl over the ground they constantly stroke the surface with the antennae.

The great brilliance and prominence of the eyes of these wasps add much to their appearance of animation and intelligence. Either the eyes are very efficient organs or else the kinesthetic sense is well developed, for when I try to "chase" a mutillid into a bottle with

THE MUTILLIDS OR COW-KILLERS

my stick she seems to sense its motion immediately and sets up the anger buzz and tries to hide. Escape is often attempted by digging into the soil, and burial accomplished in a remarkably short time. This ability to dig into the earth quickly is of high importance to the female mutillid, for it is by digging that she exposes the cells of certain host wasps and finds a place to lay her eggs. In a general way it is known that these insects are parasites on aculeate Hymenoptera (slender-waisted wasps) but few observers if any know much about the exact hosts. The adult female searches out a cocoon of a bee or wasp and lays an egg on the pupa within the cocoon. The egg hatches and the young larval mutillid feeds upon the pupa of the bee or wasp until it is full grown, and then pupates within the cocoon as the original inhabitant. At the close of the pupal period it emerges as an adult mutillid wasp.

Comparatively recently a very remarkable fact has been discovered concerning two species of African mutillids. They are parasitic on the Tsetse fly, which, you will recall, is the fly that acts as a disseminator of the organism causing the dreaded sleeping sickness.

In making her toilet this wasp is most fastidious. Brushing and combing of the hair are frequently engaged in. The antennae are cleaned by bringing the fore legs down over them. The same legs are used in combing the head parts and upper thorax. The abdomen is brushed with the rear legs.

DENIZENS OF THE MOUNTAINS

I can find no one who will venture to tell me, and I am unable to bring forth many facts myself, concerning the feeding habits of the mutillids. I have seen the winged males feeding on flowers but I have never yet been able to catch a feeding individual of the opposite sex. The female seems always to be on the move and so restless that it would appear that they never eat at all. Doubtless they do eat, and someday we will see them in the act. Let us all keep an observant eye on them.

There are many species of mutillids other than those commonly called velvet ants or cow-killers. Some of them are nocturnal and may be taken in great numbers about lights in the spring and summer. It is a very curious thing though that in the case of some of them we are able to collect only male specimens. Are the females really so rare as this would lead us to think, or is it that we have not discovered them because we do not know where they live? The last supposition is probably the true one.

INDEX

Allen, J. A. 38
Anthony gray squirrel 44

Belding ground squirrel 85
Blue-fronted jay 94
Burro, inquisitiveness of 140

Charleston Peak 80
Chickadee
    Bailey mountain 41
    devotion of female 42
    Gambel 41
    long-tailed 41
    movements 40
    nest sites 42
    song 39
    vertical distribution 41
Chipmunk
    Arizona 26
    distribution 17
    foot of 21
    Gila 18
    habitat 18
    hibernation 19
    Merriam
        economic value 38
        enemies 36
        food 37
        home 33, 37
        markings 35
        play 34
        relation to San Bernardino chipmunk 33
        young 37

INDEX

    zonal distribution...............................32
sage-bush..18
  San Bernardino
    feeding habits.........................25, 28, 29
    food..27
    hibernation...................................23
    habits..21
    nest..23
    notes...29
    parental attachment...........................24
    young...23
Cinclus...117
Cony
    bleating...................................12, 14
    description.................................9, 11
    dwelling......................................13
    enemies.......................................15
    habitat..9
    hay-stacks....................................13
    hibernation...................................16
    inquisitiveness...............................14
    protective coloring...........................15
    young...16
Cow-killer..154

Eutamias..................................17, 19, 26

Fox, gray
    bark....................................105, 110
    enemies......................................111
    food...109
    hunting.................................105, 109
    intelligence.................................108
    play of young................................106
    pets...106

INDEX

    relation to birds............................109
    young......................................105

Gale, Dennis............................42, 120
Golden-mantled ground squirrel..............122
Gray fox......................................104

Jay, blue-fronted
    description.................................98
    food.......................................100
    habitat....................................101
    mating season..............................101
    nest.......................................101
    relation to other birds................94, 99
Junco
    nest..................................147, 149
    note..................................145, 150
    occurrence.................................147
    relation to snakes....................147, 149
    Sierra.....................................145
    song.......................................145

Merriam, C. Hart..............................38
Merriam chipmunk..............................31
Mountain weasel...............................56
Mountain wood rat..............................1
Mouse, white-footed
    cannibalism.................................70
    enemies.....................................75
    food..75
    persistence in nest building................68
    rearing of young............................70
Muir, John....................................56
Mule deer....................................103
Mutillid

INDEX

    digging habits................................159
    distribution.................................155
    metamorphosis...............................159
    note..158
    sting.......................................157
    superstitions concerning.....................154

Navajo Indians....................................157
Neotoma..8
Nutcracker
    call note....................................99
    description..................................77
    food...79
    inquisitiveness...........................77, 84
    location of nest.............................82
    migrations...................................83
    nesting duties...............................83
Nuthatch
    distribution of western......................66
    food...63
    nest cavity..................................64
    note...66
    play habits..................................62
    young..64

Ouzel
    adaptations.................................116
    food..118
    desert habitat..............................114
    migration...................................115
    nest..119

Palomar Mountains..................................47
Penthestes...41

INDEX

Peromyscus 71
Pewee 58

Rat, mountain wood
    description 4
    dwellings 4
    faunal associates 8
    food 7
    signs of 5
    trading habits 7

San Bernardino chipmunk 17
San Ignacio 153
San Jacinto Mountains 31
Skunk
    beneficial rôle 139
    description 139
    enemies 144
    relation to yellow-jackets 137
    scent 142
    wanderings of male 141
    young 140, 142
Squirrel
    Anthony gray
        egg-pilfering 49
        feeding habits 53
        food 53
        nests 49
        relation to jays 47
        relation to woodpeckers 44, 47
        young, birth of 50
        young, habits of 50
    Belding ground
        call note 85
        digging habits of young 89

INDEX

    excavations...92
    food...91
    habitat...85
    young...88, 91, 93
Calico...124
Golden-mantled ground
    burrow...122
    contrasted with chipmunks...122
    distribution...123
    feeding habits...125
    food...125, 129
    hibernation...130
    swimming...129
    young...131
    Richardson red...53

Water snakes...147
Water-ouzel...112
Weasel, mountain
    distribution...61
    hunting activities...58
    nursery...60
    relation to birds...150
    relation to chipmunks...60
    temper...60
    type locality...61
    young...60
Wood rat, mountain...1
Wren-tit
    call notes...131
    classification...136
    description...131, 134
    distribution...135
    incubation...134
    nesting habits...133

THIS BOOK

DENIZENS OF THE MOUNTAINS

WAS SET, PRINTED AND BOUND BY THE COLLEGIATE PRESS OF MENASHA, WISCONSIN. THE COVER DESIGN IS BY THE DECORATIVE DESIGNERS OF CHATHAM, NEW JERSEY. THE TYPE FACE IS 11 ON 13 POINT CASLON OLDE STYLE NO. 337. THE TYPE PAGE IS 22 X 34 PICAS. THE TEXT PAPER IS 80 LB. WHITE POLYCHROME ENAMEL. THE END SHEETS ARE 25 X 38—80 LILAC LAID BUCKEYE TEXT. THE JACKET IS 25 X 38—80 LILAC LAID BUCKEYE TEXT. THE BINDING IS HOLLISTON'S WAVERLY VELLUM FINISH 306.

WITH *THOMAS BOOKS* CAREFUL ATTENTION IS GIVEN TO ALL DETAILS OF MANUFACTURING AND DESIGN. IT IS THE PUBLISHER'S DESIRE TO PRESENT BOOKS THAT ARE SATISFACTORY AS TO THEIR PHYSICAL QUALITIES AND ARTISTIC POSSIBILITIES AND APPROPRIATE FOR THEIR PARTICULAR USE. *THOMAS BOOKS* WILL BE TRUE TO THOSE LAWS OF QUALITY THAT ASSURE A GOOD NAME AND GOOD WILL.

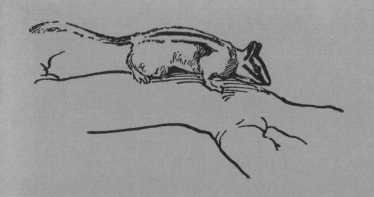